Organic Farming: Its Potential, Principles and Procedures

The Authors

Dr. Nilufar Haque has been working as Assistant Director of Extension in Directorate of Research, Extension and Farms, West Bengal University of Animal and Fishery Sciences, Kolkata, West Bengal, since July 2019. She holds master degree and Ph.D. in discipline of Animal Physiology from National Dairy Research Institute, Karnal, Haryana. She received DST inspire fellowship for her Ph.D. degree. Dr Haque has also served as Assistant Professor in Department of Physiology and Biochemistry, SDAU, Gujarat for 7 years. Dr Haque has been actively engaged in teaching and research activities. She has published about 36 articles indifferent national and international journals. She is life time member of various scientific communities such as SAPI, ANSI, WBVC, VCI etc. Dr Haque has been conferred with awards like Young Scientist Award, Best Oral Presentation Award, National level Essay Writing Competition Award. She has authored an objective question bank on Animal Physiology entitle "MCQ in Veterinary Physiology". She has also edited and compiled a book "Physio-biochemical basis of disease diagnosis".

Dr. Sheikh Asraf Hossain is working as Veterinary Officer under ARD department of West Bengal Govt. since August, 2019. Earlier, he had served as Scientist-II at Animal Nutrition Group of National Dairy Development Board (NDDB), Anand, Gujarat, India for about 7 years. He did BVSc&AH from WBUAFS, Kolkata; Post graduate (MVSc in Animal Nutrition) from AAU, Anand, Gujarat and Doctorate (Animal Nutrition) from NDRI, Karnal with merit certificate being the First class First (Divisional). He has published about 34 articles in different national and international journals. He is life time member of various scientific communities such as IDA, ANSI, ANA, WBVC, VCI etc. He is reviewer of several reputed and esteemed national and international journals. Dr Hossain is also recipient of awards like Young Scientist Award, Best Oral Presentation Award, Best Poster Award. Apart from this, he is dedicated for the upliftment of the society of the country by providing training and extension activities to the veterinarians, dairy farmers, milk union personnels, trainers, officers and administrators.

Dr. Rajesh Kumar is currently working as an Assistant Professor in the Department of Veterinary Gynaecology and Obstetrics, CVSc & AH, ANDUAT, Ayodhya, UP. Earlier, Dr Kumar has served for more than four years as VMO, department of Animal Husbandry, Uttar Pradesh. Dr Kumar hold BVSc& AH degree from CVSc & AH, ANDUAT, Ayodhya, UP and MVSc (Vety. Gynaecology & Obstetrics) from AAU, Anand, Gujarat. Dr Kumar has been awarded as young scientist, young teacher and best clinician from different societies. He has published many research papers, case reports and review articles in many national and international journals and edited/authored book entitled "Multiple Choice Questions in Veterinary Science" and "Adhunik Pashu Prabandhan and Matsya Palan Drishtika". Dr Kumar is a regular reviewer in many international and national journals. Dr Kumar also act as "member, Scientific Advisory Board" in IJLR; "Fellow member" of SAS Society, Assam; "Member", National Editorial Advisory Board in JAFST, Delhi. Dr Kumar is life member of ISSAR, SAID, S&T SIRI, SVSBT, and Veterinary Council of India.

Organic Farming: Its Potential, Principles and Procedures

– Authors –

Nilufar Haque

Assistant Director of Extension
Directorate of Research, Extension and Farm
WBUAFS, Kolkata-700037, West Bengal

Sheikh Asraf Hossain

VO
Additional Block Animal Health Centre (ABAHC)
Chorolmoni, Chanchal-II, Malda-732139, West Bengal

Rajesh Kumar

Assistant Professor
Deparment of Veterinary Gynaecology and Obstetrics
ANDUAT, Ayodhya-224229, Uttar Pradesh

BIOTECH

2022

BIOTECH BOOKS®

ISBN 978-81-7622-502-1

Published by: **BIOTECH BOOKS®**
4762-63/23, Ansari Road, Darya Ganj,
New Delhi - 110 002
Phone: 09899295259, 011-43003222
E-mail: biotechbooks@yahoo.co.in
Website: www.biotechbooks.in

Typeset at:**Classic Computer Services**
Delhi - 110 035

Printed at:**Replika Press Pvt. Ltd.**

PRINTED IN INDIA

Preface

The birthright of all living things is healthiness. This is true for soil, plant, animal and man: the health of those four is one linked chain. Organic agriculture aims at the human welfare without any destruction to the environment which is the base of human life itself. In the world where number of people is growing rapidly doubling in last 50 years, agriculture plays a noteworthy role. The structure of our agriculture based on the traditional information and practices handed down from generation to generation could not yield enough to feed the increasing population. With green revolution, where agriculture became profoundly commercialized and chemicals were introduced, agriculture made an excessive improvement in terms of efficiency and effectiveness. However, the success was at the cost of ecology and environment and to the impairment of the well-being of the people. In resource and income-poor countries, organic farming seems to offer a suitable and low-cost way to upsurge productivity and to improve farm livelihood.

Organic agriculture is challenged to reveal its potential: both as a role model and a real pathway to sustainability in agriculture and food systems. The thought of eco-functional intensification goes far beyond the limiting use of fertilizers and pesticides. It needs a fundamental redesign of farms and fields, and entails more co-operation within the organic sector. Accordingly, livestock needs to be integrated into the nutrient and organic matter circuits in order to develop the robustness and resilience of both crops and animals, with the selection of well-adapted varieties and breeds. Thus, progress in organic agriculture is a challenge for scientists, farm advisors and farmers, and desires to be addressed by refining education and by empowering participation and inter-disciplinary research.

The book is the result of continuous rigorous efforts of reading, compiling of lot of authentic sources on the subject. Every effort has been made to ensure that materials in the book are accurate, relevant and up-to-date. This book gives an overview of the principles and procedures of organic farming, history, present status and an insight into the future avenues, beneficial and ill effects of the organic farming system being felt in India, certification process, and ultimately, the efforts made by the government to promote organic agriculture in the country. This book is intended not only for organic producers but also for students, teachers of agriculture and above all, for those consumers who are in dilemma to buy organic products in high price with having little knowledge about the certification process.

We would like to pay our respect to all teachers and researchers of the past, whose work helped us in preparation of this book. We shall ever, remain thankfully indebted to all those learned souls, our former teachers, known and unknown hands, who directly or indirectly motivated us in preparation of this book. We apologize for the faux pass of the person who have extended the help in a way or other and deserve such thanks. Lastly, the authors' wishes to express his appreciation for the understanding and forgiveness of his family and children for the many times they were in absentia and other times they might better have been.

Authors will highly oblige for comments and suggestions of students, friends, colleagues and teachers for further improvement of the book in future editions.

Nilufar Haque

Sheikh Asraf Hossain

Rajesh Kumar

Contents

Chapter 1

Introduction

For the centuries, agriculture was one of the biggest contributors for the development of countries, providing revenues that enabled industrialization and poverty alleviation. After the Second World War, together with increasing global need for food, conventional agriculture took growing trend. Companies producing chemicals for military purposes lost their markets; therefore they turned to new customers – farmers. The development of agriculture was a success in terms of meeting the food need. With green revolution, where agriculture became heavily commercialized and chemicals were introduced, agriculture made a great advancement in terms of efficiency and effectiveness. In the last 40 years, agricultural production doubled even though cropland area has increased only by 12 percent. At the same time, there was 700 per cent increase in global fertilizer use and 70 per cent increase in irrigated cropland area. It provides agricultural goods for human needs but often in expense of degrading environmental conditions. Agricultural land cover 40 per cent of the terrestrial earth surface, therefore the methods chosen for agriculture are making a great impact on the environment and natural life.

Conventional Farming: Influence on Biota and Environment

Around 40 per cent of agricultural land has been strongly degraded over past 50 years by intensive use of fertilizers. By 2050, world food demand is expected to double in comparison with the levels of 20th century. Most of the fertile lands of the world are already used for cultivation; there are very few opportunities on expansion. Degrading land, increasing demand for agricultural products and biofuels caused by growing human population, changing habits and growing income will put even more pressure. Conversion of land is resulting extinction of most plant species and of animals whose habitat is determined by that plant species. Below ground organisms are also affected by land use change. Additionally, converting land to agriculture leads to local extinctions of biota in those areas, but in addition to loss of habitat, agricultural practices might also have other indirect negative effects on biodiversity. The insecticides and herbicides kill non-target species, fertilizers run off into coastal waters, removing, altering or destroying natural habitat. Fragmentation of vegetation, reduction of areas with some type of vegetation and expansion of others, introduced species are the few of many worrying consequences of agriculture that may lead loss of biodiversity.

In 2010, agriculture counted for 24 per cent of global green gas emissions. The number includes agriculture production, fertilizer use, tractors and fertilizer production. Another 11 per cent was contributed by land use change, which primarily driven by agriculture. Greenhouse gasses leads to global warming, smog and acid rains. Agriculture may also impact human health. It may happen through contact with agricultural chemicals, air, water and soil pollution as well as residues in food products. The pesticides used heavily in industrial agriculture are associated with elevated risks for workers as well as consumers; there are also links with endocrine disruption and reproductive dysfunction, as well as disruption in immune and nervous systems. Growth promoting antibiotics used in animals are thought to be driving the increase of antibiotic resistance in human. The United Nations has estimated that about 2 million poisonings and 10,000 deaths occur each year from pesticides. Many pesticides are not tested on their effects on human health.

Besides, conventional agriculture has some unavoidable negative impact on ecosystems. Several of these ecosystems provide services essential for agriculture itself. Soil formation, nutrient cycle, pollination of crops, regulation of agricultural pests, water purification and climate regulation are examples of ecosystems vital to agriculture. To keep resilience and stability of farming systems, agriculture must have management strategies that do not cause negative impact on the environment. Sustainable management incorporates different components of farming system (plants, soils, insects, fungi, animals and water) recognizing importance of their interconnections and functional relationships which means that agricultural systems rich in biodiversity are able to recover more readily from biotic and abiotic stresses such as drought, environmental degradation, pests, diseases, epidemic among the others.

Therefore, in one hand, conventional agriculture is feeding all and employing many, helping or even being the one of the main sources of income for economies of many countries, while in the other hand, it came with the price, causing number of problems few of which are depletion of natural resources, environmental degradation, impacts on human health. There are alternatives for conventional agriculture. One of which is organic agriculture which reduces negative impacts of agriculture and offers price premium for the growers. Although number of organic growers is increasing it is still very small portion of the whole agriculture.

Emergence of Organic Farming

Organic farming is one of the several approaches found to meet the objectives of sustainable agriculture. It is a natural production management system which enhances biodiversity, soil biological activities and biological cycles. It prohibits the use of synthetic inputs, such as drugs, fertilizers and pesticides. Certified organic product can bring price premium for farmer. Research made on potential of organic farming to become main source of food indicates that organic methods could produce enough food in global scale per capita to sustain current or even larger population without increasing the agricultural land base. This is because although yields of organic farms are less than of conventional in developed countries, in developing ones organic systems produce more than conventional farms. Organic farming require more labor, but they are balanced out by lower input costs and higher margins. Even usage of organic agriculture management methods are increasing globally, increase is rather slow. The small percentage among all the agricultural land with all the benefits of organic agriculture and threats of conventional one brings focus to the reasons why adoption of organic farming is not gaining momentum.

Concept of Organic Farming

Organic farming, a system based on philosophical assumptions about how nature works. It is a thoughtful and a self-conscious practice. It is allied with sustainable agriculture in the past (because it drew inspiration from the past), but it is a relatively new movement in human history, born out of a reaction to the perceived failures of industrial agriculture. Organic farming turned the gaze of millions of people, institutions, and governments backwards to look at the past as a source of guidance and wisdom. It changed how farmers over large swaths of the earth treat soil, and the crops grown on the soil. It changed how birds feed on crops, and the quality of water that drains into our streams, rivers, and oceans. It changed consumer habits, and in many ways the ethics, demanding that we think about how our food is grown and the effect our daily habits have on nature.

Definitions of Organic Farming

According to the Food and Agriculture Organization (FAO), sustainable agriculture "is the successful management of resources for agriculture to satisfy changing human needs while maintaining or enhancing the quality of environment and conserving natural resources". In philosophical terms organic farming means "farming in spirits of organic relationship". In this system everything is connected with everything else. Since organic farming means placing farming on integral relationship, we should be well aware about the relationship between the soil, water and plants, between soil-soil microbes and waste products, between the vegetable kingdom and the animal kingdom of which the apex animal is the human being, between agriculture and forestry, between soil, water and atmosphere *etc.* It is the totality of these relationships that is the bed rock of organic farming.

Now, organic farming can be defined as a system of agriculture that relies on ecological processes to minimize or eliminate the need for chemical fertilizers and pesticides in the production of crops. There are several definitions proposed by different scientific organizations. The International Federation of Organic Agriculture Movements (IFOAM, 2009) defines organic agriculture as follows: *"Organic agriculture is a production system that sustains the health of soils, ecosystems and people. It relies on ecological processes, biodiversity and cycles adapted to local conditions, rather than the use of inputs with adverse effects. Organic agriculture combines tradition, innovation and science to benefit the shared environment and promote fair relationships and a good quality of life for all involved."*

According to US Department of Agriculture (USDA, organic farming is *'a system that is designed and maintained to produce agricultural products by the use of methods and substances that maintain the integrity of organic agricultural products until they reach the consumer.*

In another definition FAO suggested that *"Organic agriculture is a unique production management system which promotes and enhances agro-ecosystem health, including biodiversity, biological cycles and soil biological activity, and this is accomplished by using on-farm agronomic, biological and mechanical methods in exclusion of all synthetic off-farm inputs"*. This is accomplished by using, where possible, on-farm agronomic, biological and mechanical methods, as opposed to using synthetic materials to fulfill any specific function within the system.

According to NPOP India *"Organic agriculture is a system of farm design and management to create an eco system, which can achieve sustainable productivity without the use of artificial external inputs such as chemical fertilizers and pesticides"*.

Organic Farming: An Environmentally Friendly Practice

The principles and methods employed in organic farming promote practices that co-exist with natural systems and help to protect and enhance the environment. From an operational perspective, it concentrates on the nourishment of the soil through the use of natural inputs, avoids the requirement for herbicides, fungicides and insecticides by using crop rotations, maximizes access to the outdoors using more appropriate breeds of animal (including traditional breeds), provides liberal space when indoors and excludes the use of Genetically Modified Organisms (GMOs).

Organic farming places a strong emphasis on environment friendly practices, with particular concern for animal welfare. Animal production is an important part of organic farming that aims at achieving a balanced relationship between the soil, the plants and the animals in a farming system. Each component is as important as the other in contributing to the overall effect and in fulfilling the key values of naturalness, harmony, local circulation of resources and the principle of precaution.

Organic farming and the organic foodstuff sector is established as a sustainable farming and production system which fulfils a dual societal role by responding to an increasing consumer demand for organic products while also delivering public goods which contribute to protection of the environment, animal welfare and rural development. Consumers expect the food from organic production to be of a certain quality that makes it different from conventionally produced food. Acknowledged consumer interests cover a wide range of issues, from the nature of farming as a whole (environment friendly, socially just, animal welfare friendly) to the concern over own health (buying organic food based on a perception of organic food being more healthy).

India is bestowed with lot of potential to produce all varieties of organic products due to its diverse agro-climatic regions. In several parts of the country, the inherited tradition of organic farming is an added advantage. This holds promise for the organic producers to tap the market which is growing steadily in the domestic market related to the export market. In India, there is considerable latent interest among farmers in conversion to organic farming. However, some farmers are reluctant to convert because of the perceived high costs and risks involved in organic farming. Despite attention which has been paid to organic farming over last few years, very little accessible information actually exists on costs and returns of organic farming in India. The empirical evidences of efficiency analysis of organic and conventional farming systems are scarce or even absent.

However, till now, organic agriculture has grown out of conscious efforts by inspired people to create best possible relationship between the earth and men. Since its beginning the sphere surrounding organic agriculture has become considerably more complex. A major challenge today is certainly its entry into the policy making arena, its entry into anonymous global market and transformation of organic products into commodities. During last two decades, there has also been a significant sensitization of the global community towards environmental preservation and assuring of food quality. Ardent promoters of organic farming consider that it can meet both these demands and become mean for complete development of rural areas. After almost a century of development organic agriculture is now being embraced by the mainstream and shows great promise commercially, socially and environmentally. While there is continuum of thought from earlier days to the present, the modern organic movement is radically different from its original form. It now has environmental sustainability at its core in addition to the founders concerns for healthy soil, healthy food and healthy people.

Principles of Organic Farming

Organic farming works in harmony with nature rather than against it. This involves using techniques to achieve good crop yields without harming the natural environment or the people who live and work in it. Organic farming does not mean going 'back' to traditional methods. Many farming methods used in the past are still useful today. Organic farming takes best of these and combines them with modern scientific knowledge. Organic farmers do not leave their farms to be taken over by nature; they use all the knowledge, techniques and materials available to work with nature. In this way the farmer creates a healthy balance between nature and farming, where crops and animals can grow and thrive.

Organic farming reflects the emphasis of stimulating and enhancing self-regulatory processes and intensification of agricultural systems through 'biological' (living organisms) and 'ecological' tools (agricultural management of the ecosystem, habitat diversity) rather than external inputs. These ideas are expressed in the four fundamental principles of organic farming – health, ecology, fairness and care (IFOAM, 2005).

1) Principle of Health

Organic Agriculture should sustain and enhance the health of soil, plant, animal, human and planet as one and indivisible. This principle points out that the health of individuals and communities cannot be separated from the health of ecosystems - healthy soils produce healthy crops that foster the health of animals and people.

Health is the wholeness and integrity of living systems. It is not simply the absence of illness, but the maintenance of physical, mental, social and ecological well-being. Immunity, resilience and regeneration are key characteristics of health. The role of organic agriculture, whether in farming, processing, distribution, or consumption, is to sustain and enhance the health of ecosystems and organisms from the smallest in the soil to human beings. In particular, organic agriculture is intended to produce high quality, nutritious food that contributes to preventive health care and well-being.

In view of this it should avoid the use of fertilizers, pesticides, animal drugs and food additives that may have adverse health effects.

2) Principle of Ecology

Organic agriculture should be based on living ecological systems and cycles, work with them, emulate them and help to sustain them. This principle roots organic agriculture within living ecological systems. It states that production is to be based on ecological processes, and recycling. Nourishment and well-being are achieved through the ecology of the specific production environment. In the case of crops this is the living soil; in animals it is the farm ecosystem; for fish and marine organisms, it is the aquatic environment. Organic farming, pastoral and wild harvest systems should fit the cycles and ecological balances in nature. These cycles are universal but their operation is site-specific. Organic management must be adapted to local conditions, ecology, culture and scale. Inputs should be reduced by reuse, recycling and efficient management of materials and energy in order to maintain and improve environmental quality and conserve resources. Organic agriculture should attain ecological balance through the design of farming systems, establishment of habitats and maintenance of genetic and agricultural diversity.

Those who produce, process, trade, or consume organic products should protect and benefit the common environment including landscapes, climate, habitats, biodiversity, air and water.

3) Principle of Fairness

Organic agriculture should build on relationships that ensure fairness with regard to the common environment and life opportunities. Fairness is characterized by equity, respect, justice and stewardship of the shared world, both among people and in their relations to other living beings. This principle emphasizes that those involved in organic agriculture should conduct human relationships in a manner that ensures fairness at all levels and to all parties - farmers, workers, processors, distributors, traders and consumers. Organic agriculture should provide everyone involved with a good quality of life, and contribute to food sovereignty and reduction of poverty. It aims to produce a sufficient supply of good quality food and other products. This principle insists that animals should be provided with the conditions and opportunities of life that accord with their physiology, natural behavior and wellbeing.

Natural and environmental resources that are used for production and consumption should be managed in a way that is socially and ecologically compatible and should be held in trust for future generations. Fairness requires systems of production, distribution and trade that are open and equitable and account for real environmental and social costs.

4) Principle of Care

Organic agriculture should be managed in a precautionary and responsible manner to protect the health and well-being of current and future generations and the environment. Organic agriculture is a living and dynamic system that responds to internal and external demands and conditions. Practitioners of organic agriculture can enhance efficiency and increase productivity, but this should not be at the risk of jeopardizing health and well-being. Consequently, new technologies need to be assessed and existing methods reviewed. Given the incomplete understanding of ecosystems and agriculture, care must be taken. This principle states that precaution and responsibility are the key concerns in management, development and technology choices in organic agriculture. Science is necessary to ensure that organic agriculture is healthy, safe and ecologically sound. However, scientific knowledge alone is not sufficient. Practical experience, accumulated wisdom and traditional and indigenous knowledge offer valid solutions, tested by time. Organic agriculture should prevent significant risks by adopting appropriate technologies and rejecting unpredictable ones, such as genetic engineering. Decisions should reflect the values and needs of all who might be affected, through transparent and participatory processes.

Organic agriculture combines tradition, innovation and science to benefit the shared environment and promote fair relationships and a good quality of life for all involved.

Are Natural and Organic Products Same?

The word "organic" has become a contemporary buzz-word, often used carelessly in marketing to signify something "natural" and inherently good. "Natural" and "organic" are not interchangeable terms. In general, "natural" on a food label means that it has no artificial colors, flavors or preservatives. It does not refer to the methods or materials used to produce the food ingredients. On the contrary, "organic animal husbandry" has a very strict definition: it denotes livestock production in organic farming; that is, farming systems that adhere to the IFOAM Basic Standards for Organic Agriculture and Food Processing (IFOAM, 2000). The word "natural" is meaningless in the food world, but the word "organic" has a strict and powerful meaning in law. If a farmer describes his produce as "certified organic" it means he has gone through an arduous and to some extent costly process to meet USDA guidelines, requiring that produce to be raised using virtually no chemical pesticides, fertilizers, or synthetic ingredients. Both seeds and transplants are chemical free, and all fertilizer is organic as well.

Traditional Farming Vs Organic Farming Same

Agrochemicals have been used in a large scale only since the 1960s. Therefore, farming communities which have not been influenced by the so called "Green Revolution" automatically meet the most important criteria of organic agriculture, *i.e.* the nonuse of any chemical fertilizer, pesticides and genetically modified organisms. These systems are over the last few decades, the focus in agriculture typically shifted from mainly subsistence agriculture (for own consumption) to market production (for gaining a financial income). In many countries, the density of population increased tremendously and many traditional farming systems have been unable to meet the yield expectations of the farmers. Due to reduced fallow periods, overgrazing or exploitative cultivation, many traditionally farmed areas face severe degradation. At the same time, high yielding crop varieties have been introduced which are more prone to diseases. Organic farming tries to meet the increased needs of the growing population while not risking the long-term productivity of the farmland is referred to as "Traditional Farming". Many methods and techniques of organic agriculture have originated from various traditional farming systems all over the world. However, not all traditional systems make use of these methods, sometimes for the simple reason that they are not known in a specific reason. In addition, organic farming disposes of rather modern technologies such as the use of antagonistic microbes in pest management, high yielding but disease resistance varieties or the use of highly efficient green manure plants. Whether a certain traditional farming system can be called organic will depend on whether all the organic standards are fulfilled. For instance, some traditional systems get in conflict with the requirements of organic animal husbandry (sufficient space and free movement), the necessary prevention of soil erosion, the ban to cut forests and to burn biomass (*e.g.* slash and burn systems).

The following table gives the comparative idea of traditional and organic farming:

What traditional and organic farming have in common	Organic methods which can be found in traditional farming	What is specific to organic farming
No use of chemical fertilizers, insecticides, fungicides, herbicides, growth promoters *etc.*	Closed nutrient cycles, low external inputs	Use of microbial preparations for pest management
No use of genetically engineered plants and animals	Recycling of biomass through mulching or composting	Release of efficient attraction of beneficial insects
Use of animal manures	Mixed cropping and/or crop rotation	Use of high yielding, but disease resistance varieties/breeds of crops/animals
	Sustainable manage or resources: soil, energy, water	Introduction of efficient green manures, cover crops and nitrogen fixing trees
	Maintenance of soil fertility	Use of improved tools for soil cultivation, weeding, sawing *etc.*
	Animal friendly husbandry practices	Application of improved compost methods and biofertilizers

Chapter 2

History of Organic Farming

Organic farming has developed from a sub-cultural protest movement in the 1970's to a more or less established part of contemporary farming. The origin of today's organic farming can be traced back along several different lines; some which originated more than a century before the turbulent days of the 1970's when the current organic farming movement started. The organic movement has decades of experience through practicing ecologically sound agriculture and also in establishing inspection and certification schemes to give the consumers the guarantee and confidence in actuality.

Today's organic farming grew out of the work of early pioneers combined with new and radical ideas flourishing at this time. The main trunk of its growth occurred in Britain and its imperial outposts in Asia, Africa, and North America, with offshoots and allied impulses in Europe, Japan, and most parts of the globe, until by twenty-first century organic farming movement has touched every region on earth.

The Early Pioneers

Most of early efforts focused on plant production and concern with human health. During first half of the 20th century pioneer researchers as well as farmers were working to develop "natural" methods to enhance soil productivity. The very beginning of organic farming may be when **Justus von Liebig (1803-1873)** presented his revolutionary theories about importance of minerals in plant nutrition. His ideas that inorganic nutrients could replace manure laid foundation for chemical use in conventional agriculture – and made it possible to develop an alternative. Less known is that von Liebig also advocated recycling of nutrients from urban areas back to fields, the use of green manure, and a balanced crop rotation, in a way that would much please agroecologists of today.

Rudolf Steiner (1924), founder of "biodynamic approach to agriculture" in Austria, saw animals as a very important part of the "farm organism" (the other parts being soil, plants and humans). He thus considered animals a natural part of every biodynamic farm. Steiner's ideas regarding the importance of allowing animals a natural behavior has lead to husbandry practices close to those advocated by modern ethologists.

Among many pioneers of organic farming, **Albert Howard (1873-1947)** and **Lady Eve Balfour (1898-1990)** were some of most influential in the English-speaking world. The British agricultural scientist Sir Albert Howard is often referred to as the father of modern organic agriculture. From 1905 to 1924, he worked as an agricultural adviser in Pusa, Bengal, where he documented traditional Indian farming practices and came to regard them as superior to his conventional agriculture science. Howard worked in both India and England during the first half of the 20th century. He came to believe that there is a connection between a healthy soil and the health of plants, animals and humans, and that the key to a healthy soil is livestock manure and composting. This view was shared by many of early critics of conventional farming. Albert Howard wrote extensively about Indian traditional agricultural practice, 'Indore method of composting' in his book 'An Agriculture Testament' in 1943.

Howard inspired Eve Balfour to start an experimental farm where his ideas about cropping could be tested in practice. She was inspired by the "Natürlicher Landbau" movement and saw agriculture and handling of soil, animals and farmers/farm workers as parts in a complex of related issues. Together with a friend, Elisabeth Tamm, she presented her ideas in a small book 'Peace with the Earth' (Swedish: Fred med jorden4). In the book she also discussed farm animal living conditions, advocating that animal husbandry should be performed according to "biological principles". Her views were shared by other radical women of the time and were included in the program of the international "Women's Organization for World Order" in 1937.

The term organic farming was coined by **Walter Ernest Christopher James, (**known by the name **Lord Northbourne)** in his book 'Look to the Land' (written in 1939, published 1940). He defined organic farm as "the farm itself must have a biological completeness; it must be a living entity, it must be a unit which has within itself a balanced organic life". From his conception of "the farm as organism," he described a holistic, ecologically balanced approach to farming.

During the 1950s, sustainable agriculture was a topic of scientific interest, but research tended to concentrate on developing the new chemical approaches. In the US, **J.I. Rodale** began to popularize the term and methods of organic growing, particularly to consumers through promotion of organic gardening.

In **1962, Rachel Carson**, a prominent scientist and naturalist, published Silent Spring, chronicling the effects of DDT and other pesticides on the environment. A bestseller in many countries, including the US, and widely read around the world, Silent Spring is widely considered as being a key factor in the US government's 1972 banning of DDT. The book and its author are often credited with launching the worldwide environmental movement.

Agriculture Before the Industrial Age

Agriculture is the cultivation of fields. The word "agriculture" is the conjunction of two Latin words, ager (field) and colere (cultivate), and suggests human skill, culture, and industry. In the contemporary age, it also connotes scientific expertise, yet still retains overtones of religious feeling and national vitality.

Agriculture embraces both crops and livestock; Humans, before industrial agriculture, did not always practice sustainable agriculture—that is agriculture practiced in a manner that did not deplete the soil and enabled more or less permanent cultivation. Egyptian peasants did not aerate the soil, nor did they apply fertilizer and dung, or rotate crops to raise the fertility. Thus, the common error that before the industrial age almost all farmers practiced sustainable agriculture, or organic farming by default, cannot be maintained.

Ancient farmers, except in a number of river systems, practiced slash and burn, or if they had no shortage of land, simply moved from plot to plot, in search of fresh soil fertility (an approach practiced by both nomad and pastoral settlers). These approaches do not all fall under "sustainable" practices because they did not always maintain soil fertility over long periods.

In India, Indians burned dried cattle dung for fuel leaving few nutrients returned to the land. Because villagers did not compost human waste and return it to the fields, British imperial officials noticed a ring of verdant growth around villages where villagers deposited sewage. Before industrial fertilizers farmers in India survived by planting varieties of wheat and other crops that flourished with low nutrients. Some may consider this a variant of sustainable farming, but one cannot describe it as organic.

In Greece and Rome free citizens cultivated their own fields, rented to tenants, hired labor, or owned slaves to work the soil for them. As wealth collected at the top of Roman society, absentee landlords and sheep farming replaced many of productive small farms, a process hastened by imports of cheap grain and other vegetable and fruit products, resulting fall in trace minerals in fruits and vegetables. It is not clear whether these changes are due to varietal differences, depletion of nutrients in the soil, yield and dry matter content differences, or diminished soil microbiological activity needed to 'unlock' nutrients from the soil and make them available to crops.

Developmental Era of Organic Farming

The concept of organic farming was started 1,000 years back when ancient farmers started cultivation near the river belt depending on natural resources only. There is brief mention of several organic inputs in Indian ancient literature like Rig-Veda, Ramayana, Mahabharata and Kautilya Arthasashthra *etc.* In fact, organic agriculture has its roots in traditional farming practices that evolved in countless villages and farming communities over the millennium. The development of the organic farming era worldwide had gone through mainly three stages, Emergence, Development, and Growth in chronological sequence.

i) Era of Emergence (1924–1970)

The beginning of organic farming could trace back to 1924 in Germany with Rudolf Steiner's course on Social Scientific Basis of Agricultural Development, in which his theory considered the human being as part and parcel of a cosmic equilibrium that he/she must understand in order to live in harmony with the environment. Organic farming received social significance in the late 1960s and early 1970s, through the "Back to Landers" movement. In this movement, the younger generation migrated from cities to rural areas – the countryside- to experiment with a utopian lifestyle. Therefore, in this era, the world started to believe that, a balance must be struck between the spiritual and material side of life.

ii) Era of Development (1970–1990)

The research and practice of organic agriculture expanded worldwide after the 1960s. In particular, the expansion and dual polarity of organic agriculture started with the oil crisis of 1973 and the growing sensitivity to agro-ecological issues. This was a time of new ideas, significant sociological transformations, protest movements and the proliferation of alternative life styles. The new thoughts in terms of using natural resources rationally, protecting the environment, realizing low input and high efficiency, ensuring food security, returning to the earth and maintaining a sustainable development of agriculture, such as organic, organic-biological, biodynamic, ecological, and natural agriculture were remarkably developed. Several organizations were established and played an important role in standardizing the production and market of organic products and promoting research and consumer's awareness. The legislative action on organic farming started gradually in the different countries and regions as the guidelines for organic farming.

In the 1970s, global movements concerned with pollution and the environment increased their focus on organic farming. As the distinction between organic and conventional food became clearer, one goal of the organic movement was to encourage consumption of locally grown food, which was promoted through slogans like "Know Your Farmer, Know Your Food". In 1972, the International Federation of Organic Agriculture Movements (IFOAM) was founded in Versailles, France and dedicated to the diffusion and exchange of information on the principles and practices of organic agriculture of all schools and across national and linguistic boundaries.

The United States Department of Agriculture (USDA) in 1980 made an investigation on a large scale on organic farming in 69 organic farms of 23 states and published the Report and Recommendations on Organic Farming, in which development status and potential remained as issues and research directions were analyzed. In this report definition and guideline for organic farming were given, and an action plan for development of organic farming was called for. The publication of this report was a milestone in legislation and development of organic farming in the United States (USDA 1980).

In the 1980s, around the world, farming and consumer groups began seriously pressuring for government regulation of organic production. This led to legislation and certification standards being enacted through the 1990s and to date.

iii) Era of Growth (Since 1990)

The rapid growth of organic farming at global scale started during end part of twentieth century, several trade organizations were founded, regulations were implemented and movements were promoted by both governmental and nongovernmental organizations. All these organizations played a pivotal role and made valiant efforts to investigate large scale organic farming with precise scientific validation. This led to rapid development of organic farming with co-ordinate and rational approach.

Organic Fair Trade Movement and Polarisation (2000 onwards)

In late 1990s, the World observed the process of globalization, industrialization and market liberalization that brought injustice to society and, ecology through manmade destruction of the world. On the other hand, these processes have created new forms of social movements in societies that are striving for creation of an economically, socially sustainable society in the World.

The Indian Experiment of Organic Farming Movement

Organic farming is very much native to this land. Whosoever tries to write a history of organic farming will have to refer India and China. The farmers of these two countries are farmers of 40 centuries and it is organic farming that sustained them. Organic agriculture began more or less simultaneously in Central Europe and India. The British botanist Sir Albert Howard, often referred to as the father of modern organic agriculture, works as an agricultural adviser in Pusa, Bengal, (now in Bihar), where he documents traditional Indian farming practices, and came to regard them as superior to his conventional agriculture science.

Before the Green Revolution, it was feared that millions of poor Indians would die of hunger in the mid 1970s. However, the Green Revolution, within a few years, showed its impact. The country, which was greatly relied on imports for its food supply, reduced its imports every passing year. As time went by, extensive dependence on chemical farming has shown its darker side. Both consumer and farmers are now gradually shifting back to organic farming in India.

From 1970s onwards, India witnessed different populist movements across the country. The country observed the genesis, growth of different farm movements such as Chipko Movement (1973), Beej Bachavo Andolan (Save Seeds Movement), which gradually became the central path for organic farming philosophy in India. The Beej Bachao Andolan has influenced practitioners to concentrate their initiatives on the conservation, protection and development of local varieties of seeds throughout the country. Beej Bachavo Andolan, a farmers' grassroots movement was influential in sowing the seeds of organic philosophy among farmers and it actively campaigned for the preservation of traditional seeds, protection of livestock, traditional knowledge and agricultural practices.

From the 1990s, civil society organizations and progressive farmers have focused on the development of different practical applications of organic farming practices through the scientific lens at the grassroots. Civil society organizations have given prominence to the development of livelihoods and environment friendly farming with localization of farming at grassroots level. Over the years, civil society organizations collaborated with organic farmers under a single banner for collective decision-making and introduced a formal association namely; OFAI (Organic Farming Association of India) in 2003 and in 2006. It was formerly registered under Societies Registration. Afterwards, understanding dearth of literature in organic and sustainable farming, development practitioners have conducted workshops, conferences and seminars across the country; brought civil society leaders, activists, organic practitioners and progressive farmers under a single platform for promoting and developing 'Indian organic agriculture'.

Chapter 3

Criteria of Organic Farming

To define the concept of organic farming, one may refer to the definition developed by the Codex Alimentarius, on the basis of contributions from experts across the globe. According to the Codex, organic farming involves holistic production management systems (for crops and livestock) emphasizing use of management practices in preference to use of off-farm inputs. This is accomplished by using, where possible, cultural, biological and mechanical methods in preference to synthetic materials.

The Codex guidelines specify that an organic production system is designed to:

- Enhance biological diversity within the whole system
- Increase soil biological activity
- Maintain long-term soil fertility
- Recycle wastes of plant and animal origin in order to return nutrients to the land, thus minimizing the use of non-renewable resources
- Rely on renewable resources in locally organized agricultural systems
- Promote the healthy use of soil, water and air as well as minimize all forms of pollution thereto that may result from agricultural practices
- Handle agricultural products with emphasis on careful processing methods in order to maintain the organic integrity and vital qualities of the product at all stages
- Become established on any existing farm through a period of conversion, the appropriate length of which is determined by site-specific factors such as the history of the land, and type of crops and livestock to be produced.

Basic Steps and Components

Organic farming approach involves management to organic management ensure biodiversity and sustainability of the system alternative sources of nutrients such as crop rotation, residue management, organic biological inputs, management of weeds and pests by better management practices, physical and cultural means and by biological control system tandem with organic concept and make them an integral part of the entire system.

Criteria for Organic Production

Organic producers are responsible for ensuring they comply with some criteria of organic regulations. The regulations are for reference to point up opportunities to work with organic producers and help for better understanding of these regulations.

i) Organic Production and Handling System Plan

The producer or handler of a production or handling operation, intending to sell, label, or represent agricultural products as "100 per cent organic," "organic," or "made with organic (specified ingredients or food group(s))" must develop an organic production or handling system plan that is agreed to by the producer or handler and an accredited certifying agent. An organic production or handling system plan must include a description of practices and procedures to be performed and maintained, including frequency with which they will be performed; a list of each substance to be used as a production or handling input, indicating its composition, source, location(s) where it will be used, and documentation of commercial availability, as applicable; a description of the monitoring practices and procedures to be performed and maintained, including the frequency with which they will be performed, to verify that the plan is effectively implemented; a description of the recordkeeping system implemented to comply with the requirements; a description of the management practices and physical barriers established to prevent commingling of organic and nonorganic products on a split operation and to prevent contact of organic production and handling operations and products with prohibited substances; and additional information deemed necessary by the certifying agent to evaluate compliance with the regulations. A producer may substitute a plan prepared to meet the requirements of another state, or local government regulatory program for the organic system plan.

ii) Land Requirements

Any field or farm parcel from which harvested crops are intended to be sold, labeled, or represented as "organic," must have had no prohibited substances, applied to it for a peiiod of 3 years immediately preceding harvest of the crop; and have distinct, defined boundaries and buffer zones such as runoff diversions to prevent unintended application of a prohibited substance to the crop or contact with a prohibited substance applied to adjoining land that is not under organic management.

iii) Soil and Crop Fertility Management

The producer must select and implement tillage and cultivation practices that maintain or improve physical, chemical, and biological condition of soil and minimize soil erosion; manage crop nutrients and soil fertility through rotations, cover crops, and application of plant and animal materials; and "manage plant and animal materials to maintain or improve soil organic matter content in a manner that does not contribute to contamination of crops, soil, or water. Raw animal manure must be incorporated into the soil not less than 120 days prior to harvest of a product whose edible portion has direct contact with the soil surface or soil particles; or incorporated into the soil not less than 90 days prior to harvest of a product whose edible portion does not have direct contact with the soil surface or soil particles. Animal manure may be applied at any time if it is composted through a process that "established an initial C:N ratio of between 25:1 and 40:1; and maintained a temperature of between 131 °F and 170 °F for 3 days using an in-vessel or static aerated pile system; or maintained a temperature of between 131 °F and 170 °F for 15 days using a windrow composting system, during which period, the materials must be turned a minimum of five times. Producers must not use sewage sludge (biosolids).

iv) Seeds and Planting Stock

The producer must use organically grown seeds, annual seedlings, and planting stock. However, when an equivalent organically produced variety is not commercially available, producers may use non-organically produced seeds. However, organically produced seed must be used for the production of edible sprouts.

v) Crop Rotation

The producer must implement a crop rotation including but not limited to sod, cover crops, green manure crops, and catch crops that provide the following functions that are applicable to the operation: maintain or improve soil organic matter content; provide for pest management in annual and perennial crops; manage deficient or excess plant nutrients; and provide erosion control. Crop rotation is defined as the "practice of alternating the annual crops grown on a specific field in a planned pattern or sequence in successive crop years so that crops of the same species or family are not grown repeatedly without interruption on the same field. Perennial cropping systems employ means such as alley cropping, intercropping, and hedgerows to introduce biological diversity in lieu of crop rotation.

vi) Crop Pest, Weed, and Disease Management

The producer must use management practices to prevent crop pests, weeds, and diseases including but not limited to crop rotation and soil and crop nutrient management practices, sanitation measures to remove disease vectors, weed seeds, and habitat for pest organisms; and cultural practices that enhance crop health, including selection of plant species and varieties with regard to suitability to site-specific conditions and resistance to prevalent pests, weeds, and diseases. Pests, weeds and diseases may be controlled through mechanical or physical methods, such as development of habitat for natural enemies of pests, mowing, and livestock grazing. If these methods are insufficient to prevent or control crop pests, weeds, and diseases, a biological or botanical substance or a substance allowed for use in organic crop production may be applied to prevent, suppress, or control pests, weeds, or diseases.

vii) Wild Crop

A wild crop represented as organic must be harvested from a designated area that has had no prohibited substance for a period of 3 years immediately preceding the harvest. A wild crop must be harvested in a manner is not destructive to the environment and will sustain the growth and production of the wild crop.

viii) Livestock Origin

Livestock products that are represented as organic must be from livestock under continuous organic management from the last third of gestation or hatching, except, poultry or edible poultry products must be from poultry that has been under continuous organic management beginning no later than the second day of life, and milk or milk products must be from animals that have been under continuous organic management beginning no later than 1 year prior to the production of the milk or milk products.

ix) Livestock Feed and Pasture

The producer must provide livestock with a total feed ration composed of agricultural products, including pasture and forage, which are organically produced and handled by operations certified as organic. The producer of an organic operation must not use animal drugs, including hormones, to promote growth; provide feed supplements or additives in amounts above those needed for adequate nutrition and health. Ruminant animals must be grazed throughout the entire grazing season for the geographical region, which shall be not less than 120 days per calendar year. Producers must provide pasture of a sufficient quality and quantity to graze and to provide ruminants with an average of not less than 30 per cent of their dry matter intake from grazing throughout the grazing season.

x) Livestock Health Care

The producer must establish and maintain preventive livestock health care practices, including selection of species and types of livestock with regard to suitability for site-specific conditions and resistance to prevalent diseases and parasites, provision of a feed ration sufficient to meet nutritional requirements, appropriate housing, pasture conditions, and sanitation practices to minimize the occurrence and spread of diseases and parasites, vaccines and other veterinary biologics. When preventive practices and veterinary biologics are inadequate to prevent sickness, a producer may administer synthetic medications. However, a producer may not use any antibiotics, any substance that contains a synthetic substance not allowed on the National List, any animal drug in the absence of illness, hormones for growth promotion or withhold medical treatment from a sick animal in an effort to preserve its organic status. All appropriate medications must be used to restore an animal to health when methods acceptable to organic production fail. Livestock treated with a prohibited substance must be clearly identified and shall not be sold, labeled, or represented as organically produced.

xi) Livestock Living Conditions

The producer of an organic livestock operation must establish and maintain year-round livestock living conditions which accommodate the health and natural behavior of animals suitable to the species, its stage of life, climate, and the environment. Including, but not limited to, year-round access for all animals to the outdoors, shade, shelter, exercise areas, fresh air, clean water for drinking, direct sunlight, and appropriate clean, dry bedding. Roughages used as bedding must be organically produced. The producer of an organic livestock operation may provide temporary confinement or shelter for an animal due to inclement weather; the animal's stage of life to protect the health, safety, or well-being of the animal; to protect soil or water quality; breeding and shipping. The producer of an organic livestock operation must manage manure in a manner that does not contribute to contamination of crops, soil, or water by plant nutrients, heavy metals, or pathogenic organisms and optimizes recycling of nutrients, and must manage pastures and other outdoor access areas in a manner that does not put soil or water quality at risk.

xii) Pasture Management

The producer must have a functioning management plan for pasture. The plan must include a description of the cultural and management practices to be used to ensure pasture of a sufficient quality and quantity is available to graze throughout grazing season; types of grazing methods to be used in the pasture system; location and types of fences; location and source of shade and water; soil fertility and seeding systems; erosion control and protection of natural wetlands and riparian areas practices. Pasture must be managed to refrain from putting soil or water quality at risk.

xiii) Organic Handling Requirements

Mechanical or biological methods, including but not limited to cooking, baking, curing, heating, drying, mixing, grinding, churning, separating, distilling, extracting, slaughtering, cutting, fermenting, eviscerating, preserving, dehydrating, freezing, chilling, or otherwise manufacturing, and the packaging, canning, jarring, or otherwise enclosing food in a container may be used to process an organically produced agricultural product for the purpose of retarding spoilage or otherwise preparing the agricultural product for market. Nonagricultural substances and non-organically produced agricultural products allowed may be used in or on a processed agricultural product intended to be sold, labeled, or represented as organic, if not commercially available in organic form.

xiv) Commingling and Contact with Prohibited Substance Prevention Practice Standard

The handler of an organic handling operation must implement measures necessary to prevent the commingling of organic and nonorganic products and protect organic products from contact with prohibited substances.

xv) Record Keeping

A certified operation must maintain records concerning the production, harvesting, and handling of agricultural products that are intended to be sold, labeled, or represented as organic. Records must "be maintained for not less than 5 years beyond their creation; and be sufficient to demonstrate compliance with the regulations.

xv) Excluded Methods: Genetically Engineered Seeds and Inputs

In addition to being free from prohibited substances, products sold as organic must be produced and processed without the use of "excluded methods." Excluded methods include genetically engineered crop seeds, seedlings, and planting stock, and biopesticides and other agricultural inputs based on or derived from genetically modified organisms (or GMOs).

The following two tables enlist the conditions permitted or restricted for organic farming

Conditions for Products Used in Fertilization and Soil Conditioning in Organic Farming

Items	*Conditions for Use*
MATERIAL FROM PLANT AND ANIMAL ORIGIN	
Matter produced on an organic farm unit	
✰ Farmyard and poultry manure, slurry, urine	Permitted
✰ Crop residues and green manure	Permitted
✰ Straw and other mulches	Permitted
✰ Compost and vermicompost	Permitted
Matter produced outside the organic farm unit	
✰ Blood meal, meat meal, bone meal and feather meal without preservatives	Restricted
✰ Compost made from plant residues and animal excrement	Restricted
✰ Farmyard manure, slurry, urine	Restricted
✰ Fish and fish products without preservatives	Restricted
✰ Guano	Restricted
✰ Human excrement	Restricted
✰ Wood, bark, sawdust, wood shavings, wood ash, wood charcoal	Restricted
✰ Straw, animal charcoal, compost and spent mushroom and vermiculate substances	Restricted
✰ Compost from organic household	Restricted
✰ Compost from plant residues	Restricted
✰ Sea weed and sea weed products	Restricted
By products from the industries	
✰ By products from the food and textile industries of biodegradable material of microbial, plant or animal origin without any synthetic additives	Restricted
✰ By products from oil palm, coconut and cocoa (including fruit bunch, palm oil mill effluent, cocoa peat and empty cocoa pods	Restricted

Items	Conditions for Use
☆ By products of industries processing ingredients from organic agriculture	Restricted
☆ Extracts from mushroom, Chlorella, fermented product from Aspergillus, natural acids (vinegar)	Restricted
Mineral origin	
☆ Basic slag	Restricted
☆ Calcerous and magnesium rock	Restricted
☆ Lime, limestone, gypsum	Permitted
☆ Calcifera sea weed	Permitted
☆ Calcium chloride	Permitted
☆ Mineral potassium with low chlorine content (eg. Sulphate of potash, kainite, sylvinite, patenkali)	Restricted
☆ Natural phosphates (rock phosphate)	Restricted
☆ Trace elements	Permitted
☆ Sulphur	Permitted
☆ Clay (bentonite, perlite, zeolite)	Permitted
Microbiological origin	
☆ Bacterial preparations (biofertilizers)	Permitted
☆ Biodynamic preparations	Permitted
☆ Plant preparations and botanical extracts	Permitted

Conditions for Products Used in Plant Pest and Disease Control

Items	Conditions for Use
MATERIAL FROM PLANT AND ANIMAL ORIGIN	
☆ Plant based repellents (neem preparations from *Azadirachta indica*)	Permitted
☆ Algal preparations (gelatin)	Permitted
☆ Casein	Permitted
☆ Extracts from mushroom, chlorella, fermented products from Aspergillus	Permitted
☆ Propolis	Restricted
☆ Beeswax, natural acids (vinegar), plant oils, Quassia	Permitted
☆ Rotenone	Restricted
☆ Tobacco tea (pure nicotine is prohibited)	Restricted
☆ Preparation from *Ryania* species	Restricted
Mineral origin	
☆ Chlorides of lime/soda	Restricted
☆ Burgundy mixture	Restricted
☆ Clay	Permitted
☆ Copper salts/inorganic salts	Not allowed
☆ Quick lime	Restricted

Items	*Conditions for Use*
Insects origin	
☆ Release of parasites, predators of insect pests	Restricted
☆ Sterilized insects	Restricted
Microorganisms used for biological pest control	
☆ Viral, fungal and bacterial preparations (biopesticides)	Restricted
Trap, barriers and repellants	
☆ Physical methods (eg chromatic traps, mechanical traps)	Permitted
☆ Mulches, nets	Permitted
☆ Pheromones – in traps and dispensers only	Permitted

Chapter 4

Techniques of Organic Farming

Environmentally Friendly Production System

Organic farming management is an integrated approach, where all aspects of farming systems are interlinked with each other and work for each other. Organic management stresses on optimization of resource use and productivity, rather than maximization of productivity and over exploitation of resources on the cost of resources meant for future generations.

Important goals of organic farming are:

- ☆ A sufficiently high level of productivity
- ☆ Compatibility of cultivation with the natural cycles of the production system as a whole
- ☆ Maintaining and increasing the long-term fertility and biological activity of the soil
- ☆ Maintaining and increasing natural diversity and agro-biodiversity
- ☆ Maximum possible use of renewable resources
- ☆ Creation of a harmonic balance between crops and animal husbandry
- ☆ Creation of conditions in which animals are kept that correspond to their natural behavior
- ☆ Protection of, and learning from, indigenous knowledge and traditional management systems

Organic farming gives importance to environmental protection. Organic farmers have to follow the norms set by the local organic farming associations. Organic gardening, including vegetable gardening, is also a part of organic farming.

The basic concepts behind organic farming are:

1) It concentrates on building up the biological fertility of the soil so that the crops take the nutrients they need from secure turnover within the

soil nutrients produced in this way and are released in harmony with the need of the plants.

2 An ecological balance is maintained within the system and by the use of bio-pesticides and various cultural techniques such as crop rotation, mixed cropping and cultivation.

3) Organic farmers recycle all wastes and manures within a farm.

In a situation where conservation of energy and resources is considered to be important, community or country would make every effort to recycles to all urban and industrial wastes back to agriculture and thus the system would be requiring only a small inputs of new resources to "Top Up" soil fertility.

Features of Organic Farming

The farming practice which involves use of eco-friendly methods to grow crops and exclusion of synthetic products, such as chemical fertilizers, insecticides and pesticides are described as organic farming. A holistic approach towards growing crops, organic farming methods helps apply simple and eco-friendly techniques in farming. Use of compost fertilizers, crop rotation and biological pest control, are some of the features of organic farming methods. The farming methods that make use of various traditional agricultural practices like minimum tillage, composting, crop rotation, biological pest control, *etc.*, and exclude the application of synthetic fertilizers, insecticides, growth regulators and genetic modification of crop species, are included in organic farming methods. The use of modern technology in combination with organic farming practices helps in creating a balanced and sustainable environment for crop growth. Organic farming methods take a integrated approach in growing crops rather than exploiting the available natural resources It is aimed at enhancing the productivity of crops without use of any kind of synthetic materials and adopting sustainable approaches towards farming.

Some features of organic farming are discussed below:

1) Mixed Farming

Monoculture is the agricultural practice of producing or growing a single crop, plant, or livestock species, variety, or breed in a field or farming system at a time. Though the motive behind cultivating a single crop in monoculture is to reduce cost incurred in fertilizers, seeds and pesticides *etc.*; however, it creates problems in the long run. The reduction in fertility of the soil owing to extraction of nutrients over a long period and soil erosion result from practice of monoculture. Moreover, the pests become resistant to pesticides.

However, in mixed method of farming, a variety of crops are cultivated on a single piece of land. It helps attract different soil microbes. Some crops act as repellents to pest and these results in pest control, in an organic manner. In organic agriculture systems, one strives for appropriate diversification, which ideally means mixed farming, or the integration of crop and livestock production on the farm. In this way, cyclic processes and interactions in the agro-ecosystem can be optimized, like using crop residues in animal husbandry and manure for crop production.

It is outstanding feature of organic farming in which variety of crops are grown simultaneously or at different time on the same land. In every season care should be taken to maintain legume cropping at least 40 per cent. Mix cropping promotes photosynthesis and avoids the competition for nutrients because different plants draw their nutrients from different depth of soil. The legume fixes atmospheric nitrogen and make available for companion or succeeding crops. Deep rooted plants draw nutrient from deeper layer of soil and bring them to the surface of soil through their leaf fall. So the nutrients leached down to lower strata are further brought back to upper layer by these deep rooted plants. It also helps in protecting soil from soil erosion.

Farmers should select the crops combination according to their needs and season. In selecting crop combinations, it is also to be kept in mind that plants also have their feelings, likes and dislike *e.g.* maize gets along well with beans and cucumber, tomatoes go well with onions and marigold. On the other hand beans and onions do not go well with each other.

Entire farm should have at least 8-10 types of crops at all the times. Each field/ plot should have at least 2-4 types of crops out of which one should be legume. In case if only one crop is taken in one plot then adjacent plots should have different crops. For maintenance of diversity and pest control randomly plant 50-150/acre vegetable seedlings for home consumption and 100 plants/acre of marigold (Genda) in all crop fields. Even high nutrient demanding crops such as sugarcane can also be grown with suitable combination of various legume and vegetable crops with optimum productivity.

2) Soil Fertility

In most part of the country, poor soil health due to loss of organic matter and soil microbial load is a major problem. Reducing water availability and increasing temperature is further adding to the problems. Too much dependence on market for supply of inputs and energy has made the agriculture a cost intensive high input enterprise with diminishing returns. We need to address all these concerns and develop a system which is not only productive and low cost but also resource conserving and sustainable for centuries to come.

Organic farming has expanded rapidly in recent years and is seen as a sustainable alternative to intensive agricultural systems, developed over the last 50 years which has limited inputs of only permitted fertilizers. There is practice of growing plants with prolific leaf growth (like alfalfa) and burying them in the soil before the cultivation of main crop. The green manuring crops add organic matter to the soil that is necessary for plant growth. A living soil is the basis of organic farming. A living soil can be maintained by continuous incorporation of crop and weed biomass, use of animal dung, urine-based manures (FYM, NADEP, vermicompost), biofertilisers and bioenhancers, special liquid formulations (like vermiwash, compost tea *etc.*) during a crop's duration.

As a thumb rule, crop residues should be returned to the plot, directly or indirectly. Cattle droppings may be returned to the field as compost. As a strategy, the quantity of biomass removed for human food and fiber, cattle feed or firewood from an organic farm should be replaced with any other bio-waste on the farm. However, it is important to account for it for preparing the balance sheet of nutrients for each crop being cultivated on the farm. In phosphorous-deficient and acidic soils, some quantity of mineral grade rock phosphate and lime can also be added either by direct application to the field or through addition to compost. The compost can be further enriched by incorporation of biofertilisers, microbial inoculants, *etc.* Special composts like biodynamic compost, cowpat pit compost *etc.* are also useful and ensure optimum productivity. Use of EM (effective microorganism) formulation has also been found effective in soil enrichment and compost making. For high nutrient demanding crops and for intermittent soil enrichment use of oilcakes, poultry manure, concentrated manures (mixture of oil cakes, poultry manure and rock phosphate) can also be an ideal low-cost option of manuring.

Application of liquid manure (for soil enrichment) is essential to maintain the activity of microorganisms and other life forms in the soil. 3-4 applications of liquid manure are essential for all types of crops. Vermiwash, compost tea, cow urine, Pachgavya and Biosol *etc.* are excellent growth promoters when used as foliar spray. 3-5 sprays after 25-30 days of sowing ensure good productivity. Use of Biodynamic preparations, such as BD-500 and BD-501 as foliar spray has also been found to be effective in growth promotion.

Trees on utility space can be allowed to grow fully. Trees and bushes on farm bunds should be placed randomly at sufficient distance and pruned at repeated intervals. Glyricidia plants should be planted at close spacing on all major bunds and all around the farm. They will act not only as biological fence but also provide biologically fixed nitrogen to soil. A 400 mt long Glyricidia strip can provide 22.5 kg N/ha per year from the year 3 and up to 77 Kg N/ha from year 7 under rainfed conditions. This can be 75-100 per cent higher under irrigated conditions. Three to four harvests can be made under irrigated conditions and two harvests under unirrigated conditions. These trees should not be allowed to grow above 5.5 ft to avoid shading effect. Lopping is used as green leaf manure.

Status of Rich and Live Organic Soil

A fertile and live organic soil ideally should have organic carbon between 0.8-1.5 per cent. At any point of time, it should have adequate quantity of dry, semi decomposed and fully decomposed organic matter for the use of micro-flora and fauna. Total microbial load (bacteria, fungi and actinomycetes) should be above 1×10^8/gm of soil. There should be at least 3-5 earth worms/cubic ft of soil with enough quantity of small life forms and insects such as ants *etc.*

a) Use of Biofertilizers and Microbial Cultures

Biofertilizers *viz.*: *Rhizobium*, *Azotobacter*, *Azospirillum* and *Pseudomonas etc.* have been found to be very effective tools of fertility management and biological nutrient mobilization. Recently customized consortia of such biofertilizer organisms, better adapted to local climatic conditions have also been developed and are available commercially. Efficiency of such microbial formulations is much higher under no-chemical use situations, therefore application of such inputs need to be ensured under all cropping situations.

Method of Application

Biofertilizers can be applied to different crops and plants by following three different ways.

1) **Seed treatment:** Nitrogen fixing and phosphate stabilizing bacteria (PSB) should be suspended in water and should be poured this slurry on seed. These treated seeds should be dry in shade and to be sown immediately. For acidic and alkaline soils, it is always advisable to use slacked lime or gypsum powder respectively for coating the wet biofertilizer treated seeds.
2) **Seedling root dip treatment:** Nitrogen fixing (Azotobacter / Azospirillum) and PSB should be suspended in water and the roots of seedlings should be dipped for 20-30 min before transplanting. In case of paddy, the roots of seedlings should be dipped for 8-12 hours (overnight) before transplant.
3) **Soil treatment**: For soil treatment depending upon the total number of plants per acre, 2-4 kg of *Azotobacter/Azospirillum* and 2-4 kg of PSB are required for one acre. These two types of biofertilizer should be mixed in 2-4 liters of water separately and should be sprinkled this suspension on two separate heaps of 50-100 kg of compost. The two heaps should be mixed separately and left for incubation overnight. After 12 hours, two heaps can be mixed together.

For acidic soils, 25 kg of lime should be mixed with this mixture. In plantation crops, this mixture should be applied at the root zones by dibbling. In sugarcane the biofertilizer manure is to be applied in furrows near the root zone, after 30-40 days of planting and covered with soil. In potato, it is to be applied after 20 days of planting or at the time of earthing-up operations. In case of sugarcane and potato, if setts/tubers are not treated with plant protection chemicals then biofertilizer compost mixture can be applied in furrows immediately before planting some important formulations for soil enrichment.

Preparation of Liquid Manures

Many variants of liquid manures are being used by farmers of different states. Few important and widely used formulations are given below:

Sanjivak –100 kg cow dung, 100 lit cow urine and 500 gm jaggary should be mixed in 300 lit of water in a 500-lit closed drum. It should be fermented for 10 days and then diluted with 20 times water and should be sprinkled in one acre either as soil spray or along with irrigation water.

Jivamrut –10 kg cow dung, 10 lit cow urine, 2 kg jaggary, 2 kg any pulse grain flour and 1 kg live forest soil should be mixed in 200 lit water and fermented for 5 to 7 days and should be stirred the solution regularly three times a day. It may be used in one acre with irrigation water.

Amritpani - 10 kg cow dung should be mixed with 500 gm honey thoroughly to form a creamy paste. 250 gm of cow desi ghee should be added to it and should be mixed at high speed. After that, it should be diluted with 200 liter water. This suspension may be sprinkled in one acre over soil or with irrigation water. After 30 days, second dose should be applied in between the row of plants or through irrigation water.

Panchgavya – 5 kg fresh cow dung, 3 lit cow urine, 2 lit cow milk, 2 lit curd, 1 kg cow butter oil should be mixed and fermented for 7 days with twice stirring per day. Then 3 lit of Panchgavya should be diluted in 100 liter water and should be sprayed over soil. 20 lit panchgavya is needed per acre for soil application along with irrigation water.

Enriched Panchgavya (or Dashagavya) –5 kg cow dung and 1 kg deshi cow ghee should be mixed in a container and fermented for 3 days with intermittent stirring. After that, cow urine 3 lit, cow milk 2 lit, curd 2 lit, sugarcane juice 3 lit, tender coconut water 3 lit, banana paste of 12 fruits toddy or grape juice 2 lit should be added on the fourth day and fermented for 15 days with stirring twice daily. The formulation will be ready in 18 days. Sugarcane juice can be replaced with 500 g jaggery in 3 liters water. In case of non-availability of toddy or grape juice, 100g yeast powder mixed with 100 g jaggery and 2 lit of warm water can also be used. For foliar spray, 3-4 lit panchgavya is diluted with 100lit water. For soil application, 50 lit panchagavya is sufficient for one hector. It can also be used for seed treatment.

b) Management of Temperature

Temperature in tropical countries like India in summer season is quite high and need to be managed. It can be achieved by keeping soil covered with biological mulch. Surface mulch has been reported to conserve soil moisture and improve water use efficiency. Temperature control can also be achieved by planting different types of trees like neem, amla, tamarind, gular, zizipus bushes, gliricidia on bunds.

c) Protection to all Life Forms

Practice of maintaining enough biomass and mulching with crop and weed residue will ensure the protection to all life forms in soil. For the survivability of different life forms, the field must have dry organic matter as a food for small insects and small animals in soil, semi decomposed organic matter as food for earthworms and fully decomposed organic matter for micro organisms in the soil at all times. These insects, small animals, earthworms and microorganisms are the tireless natural employees of the soil, wherein small animals and insects feed on the larvae of pests and thus controlling the pest. Earthworms make the soil porous thus creating the more aerobic conditions in soil and also decompose the half digested organic residue and release locked nutrients into soil. Soil rich in organic carbon contain ample quantity of beneficial micro flora which plays an important role in recycling of nutrients and nitrogen fixation, phosphate solubilization and photosynthesis activity, cellulolytic activity. Therefore protection to all life forms in soil should be ensured at all time.

3) Crop Rotation

Within the mixed farm setting, crop rotation takes place as the second principle of organic agriculture. In spite of the classical rotation involving one crop per field per season, inter cropping, mixed cropping and relay cropping are other options to optimize interactions. In addition to plant functions, other important advantages such as weed suppression, reduction in soil-borne insects and diseases, complimentary nutrient supply, nutrient catching and soil covering can be mentioned.

Crop rotation is the back bone of organic farming practices. To keep the soil healthy and to allow the natural microbial systems working, crop rotation is must. Crop rotation is the succession of different crops cultivated on same land. 3-4 years rotation plan should be followed. All high nutrient demanding crops should precede and follow legume dominated crop combination. Rotation of pest host and non pest host crops helps in controlling soil borne diseases and pest. It also helps in controlling weeds. It is better for improving productivity and fertility of soil. Crop rotations help in improving soil structure through different types of root system. Legumes should be used frequently in rotation with cereal and vegetable crops. Green manure crops should also find place in planning rotations. High nutrient demanding crops should always be followed by legume crops.

4) Development of Organic Farm

As organic management is an integrated approach, manipulation and adoption of one or few steps may not yield significant results. For optimization of productivity all the essential components need to be developed in a systematic manner. These steps include: (i) Habitat development, (ii) on-farm facilities for input production (iii) Multiple cropping and crop rotation (iv) cropping sequence and combination planning, and (v) growing of crops suiting to the region, soil and climate.

i) Habitat Development

a) *Development of Infrastructure*

3-5 per cent of farm space must be reserveed for utilities, such as space for cattle, vermicompost bed, compost tank, vermiwash/compost tea unit *etc.* 5-7 trees should be planted only on this space, as all utility infrastructure need shade. Irrigation well, water pumping infrastructure *etc.* can also be in this utility area. Some percolation tanks (7x3x3mt or of any other size depending upon the rainfall and run-off pattern) should be dig for rain water conservation (1 pit per ha) at appropriate places depending upon slope and water flow. If possible, a farm pond should be developed preferably of 20x10 meter size. Few 200 lit tanks (1 per acre) should be kept for liquid manure preparation and few containers for botanicals. For 5 acre farm, 1-2 vermicompost beds, 2 biodynamic compost beds, 2-3 compost tea/vermiwash units, 5 liquid manure tanks, five cowpat pits and one underground cattle-urine collection tank should be developed.

b) *Habitat and Biodiversity*

Management of an appropriate habitat for sustenance of different life forms is an essential component of organic farming. This can be achieved by ensuring crop diversity and by maintaining a wide variety of trees and bushes as per climatic suitability. These trees and bushes will not only ensure the nutrients from air and deep soil layers to surface layer but also attract the birds and predators, friendly insects and also provide the food and shelter. There may be some loss of productivity due to shading effect but that loss can be compensated with reduced pest problems and natural biological pest control system. In the plains, for a 10-acre farm, at least five to six neem trees (*Azadirachta indica*), one to two tamarind (*Tamarindus indica*), two gular (*Ficus glumerata*), eight to ten ber (*Zizyphus* sp.) bushes, one to two aonla (*Emblica officinalis*), one to two drumstick and 10–15 wild bushes should be planted.

More specifically, if farm areas are classified into wet and dry farms, then on the wet farms, there should be five to six neem trees, one to two wood apples, one to two star fruit, eight to ten guava or sour soap, three to four drumstick, one to two fig and 10–15 bushes of mulberry, star gooseberry, curry leaf *etc.*, and on the dry farms there must be at least five to six neem, one to two bael fruit, eight to ten ber or custard apple, one to two aonla, one to two drumstick and 10–15 bushes of sasaka, nirgundi (*Vitex negundo*), *Cassia auriculata, C. tora, etc.*

In hilly areas, *Alnus nepalensis* is considered to be a wonder tree as it fixes good amount of nitrogen. It is being promoted in a cropping system mode particularly in northeastern India. Bushes of Prunus, oak (*Quercus glauca*), Pinus species along the farm boundary and yarrow (*Achillea millifolium*), buck wheat (*Fagopyrum esculentum*), lupin (*Lupinus sativus*), Himalayan stinging nettle (*Urtica parviflora*), marigold, *etc.*, in between the plots invite a lot of predators and also attract a large number of pests.

Fruit orchards also need to maintain adequate diversity with at least 3-5 types of fruit plants and few non-fruit trees (as listed above). Major and minor plots should be separated by bunds about 1.5m wide and should be planted with Glyricidia, perennial *Sesbania* (jayanti), *Leucaena leucocephala, Cassia siamea, etc.* The internal hedgerow should consist of perennial pigeon pea, Crotalaria, seasonal Sesbania, *etc.* Lops from these trees will provide enough quantity of biologically fixed nitrogen.

In between Glyricidia/Sesbania rows few plants of pesticidal value should be inserted such as *Adathoda vesica, Vitax nigundo, Calotropis, Datura alba, Ipomea* (Besharam) *etc.* Surrounding the farm or garden, there should be hedgerows or a live fence of coppiced or pollarded, multipurpose, deep-rooted trees and shrubs and medicinal herbs such as Adathoda vasica, Vitex negundo, Jatropha curcas, *etc.* Ecological diversity is an essential component of any successful organic farming system.

ii) On-farm Facilities for Input Production

It is widely known fact that some biological processes of plants involved in acquiring nutrients such as nitrogen *e.g.* N_2 fixation are generally inhibited by adding Nitrogen fertilizer. Soil scientists generally cautious against nonjudicious fertilizer use and encourage use of organic compost otherwise it may lead to deficiency of micronutrients. Therefore in organic farming systems there is no place for chemicals.

a) Low Input Alternative

In first year simultaneously three different types of legumes should be sown in strips, first of 60 days (like moong), second of 90-120 days (Cow pea or soybean) and third of more than 120 days (red gram) in strips. Mixture of compost and vermicompost (2:1) @ 2.5 ton per acre enriched with 4 kg *Azotobacter* and 4 kg PSB biofertilizers or 4 kg consortia of customized cultures should be applied as basal dose at the time of sowing preferably in furrows below the seeds. Seeds of legumes should be treated with crop specific strains of *Rhizobium* biofertilizer. The entire surface should be mulched with a thick layer of biological mulch and the biomass should be drenched with Jivamrut @ 200 lit per acre. Seedlings will emerge from this layer. If soil is poor in phosphorus, then 300 kg of low grade mineral rock-phosphate should be applied along with the compost. Second dose of Jivamrut should be applied after 25-30 days of sowing with irrigation water or during rains.

To add to diversity, 100 plants/acre of marigold or *Hibiscus subdarifa* or any other suitable plant effective as trap crop/plant may be planted randomly throughout the field. Few seedlings of vegetables such as chillies, tomato, brinjal, *etc.* and rhizomes of turmeric, ginger *etc.* can be planted randomly for home consumption. The crop biomass may be collected at the end of strips in the form of heaps and drenched with Jivamrut. Short duration leafy vegetables (such as fenugreek or spinach) in the space vacated by the first and second crop may be sown and the surface may be mulched with treated biomass.

b) High Input Alternative

2.5-3.0 ton compost/vermicompost or 1.5 ton of biodynamic compost, 500 kg crushed oil cakes, 500 kg rock phosphate, 100 kg neem cake, 5 kg *Azotobacter* and 5kg PSB biofertilizer or 4 kg consortia of customized cultures must be incorporated in soil through broadcasting or by drilling in furrows below the seeds. 3-4 types of different crops may be sown in strips. 40 per cent crop stand should be of legumes. 100-150 marigold and vegetable seedlings may be randomly planted for increased diversity. For second crop also, similar quantities of manures should be used. Liquid manure (Jivamruta) @ 200lit/acre 3-4 times during cropping season along with irrigation water may be used. For increased productivity, 2-3 sprays of vermiwash or vermiwash+cow urine or Panchgavya can also be provided. In fruit orchards 3-4 types of legume mixtures as mixed or intercrop may be cultivated in inter spaces along with adequate quantity of manures (as specified above). After pod/grain harvest, the entire soil surface shall be mulched with the left over biomass and drench the biomass with 2 applications of Jivamruta.

After about 12-18 months the soil will be ready for organic cultivation of any crop combination. For next two-three years, along with any crop legumes may be incorporated as inter or companion crops. Crop residue always have at least 30 per cent residue from legumes.

iii. Cropping Sequence and Combination Planning

Some important benefits of crop rotations are:

a) Not all plants have same nutritive needs
b) Soil structure is improved through different types of roots

c) Pest build up is avoided

d) Rotations help against the build up of weeds.

Under Network Project on Organic Farming (NPOF of ICAR) important cropping systems, which were found economically better or at par with conventional system at different experimental stations in the country are as follows:

- ☆ Soybean - Berseem/Mustard/chickpea at Raipur, Chattisgarh
- ☆ Tomato/Cabbage – cauliflower – pea and maize – garlic at Bajaura, Himachal Pradesh
- ☆ Rice – wheat/potato/mustard/lentil at Ranchi, Jharkhand
- ☆ Groundnut – rabi Sorghum, soybean – durum wheat, potato – chick pea, chillil+ Cotton and maize – chick pea at Dharwad, Karnataka
- ☆ Soybean – durum wheat/mustard/chick pea/isabgol at Bhopal, M.P.
- ☆ Rice – durum wheat/berseem, rice – potato – Okra and rice – garlic, sorghum – berseem, maize – berseem – maize + cowpea and sorghum + cluster bean – oats-cowpea at Ludhiana, Punjab
- ☆ Maize – cotton, chillies – onion and brinjal – sunflower at Coimbatore
- ☆ Sorghum – pea – okra at Modipuram, Uttar Pradesh

Carrot/rice (pre kharif) – rice (kharif), potato/rice (pre kharif) – rice (kharif), tomato/rice (pre kharif) – rice (kharif), French bean/rice (pre kharif) – rice (kharif) at Umiam, Meghalaya

iv) Seed/Planting Material Treatment

In organic management, protection measures are used only in the case of problematic situations. Use of disease free seed stock and resistant varieties is the best option. There is no standard formulation or treatment methodology, available as on today, but farmers use different methods. Few of such innovative seed treating formulations are as follows:

- ☆ Hot water treatment at 53°C for 20-30 min.
- ☆ Cow urine or cow urine-termite mound soil paste
- ☆ Beejamrut
- ☆ Asphoetida 250gm in one lit. of water for 10 kg seed
- ☆ Turmeric rhizome powder mixed with cow urine
- ☆ Panchgavya extract
- ☆ Dashparni extract
- ☆ Trichoderma viride (4gm/kg seed) or Pseudomonas fluorscens (10gm/kg seed)
- ☆ Biofertilizers (Rhizobium/Azotobacter +PSB)

5) Pest Control

Organic farming may contribute substantially to future agricultural production worldwide by improving soil quality and pest control, thereby reducing environmental impact of conventional farming. It is an important aspect in the growth of any crop. Organic pest control involves undertaking various activities to control pests without using chemical pesticides and insecticides. The growth of beneficial insects is encouraged by growing suitable plants which attract them.

Beneficial insects are actually predators which control harmful insects. Disease resistant varieties are chosen for cultivation, in order to keep diseases at bay without having to spend money on costly pesticides. Special types of crops known as companion crops are grown to control pests. These crops help in diverting or discouraging the growth of harmful pests. Biological pesticides such as neem extract are useful in controlling many different pests. The practice of crop rotation helps in disturbing the reproduction cycles of pests, thereby inhibiting their growth and protecting the crops.

Organic farming performs better than conventional farming with regard to soil organic matter. A major objective of organic farming is to encourage a higher level of biological activity in the soil, in order to sustain its quality and thereby promote metabolic interactions between the soil and plants. Organic farming prohibits the use of pesticides and artificial fertilizers and encourages sympathetic habitat management, such as nitrogen-building leys to increase soil fertility. Organic matter has profound impacts on soil quality, such as enhancing soil structure and fertility and increasing water infiltration and storage. If the soil organic matter content drops below 3.5 per cent, the soil suffers an increased risk of erosion.

As in organic farming management use of synthetic chemicals are prohibited, the pest management is done by: (i) cultural or agronomic (ii) mechanical (iii) biological or by (iv) biopesticides (v) botanical pesticides or some chemicals such as copper sulphate and soft soap *etc.*

i) **Cultural alternative** - Use of disease free seed or stock and resistant varieties are best preventive practice in organic pest management. Maintenance of biodiversity, effective crop rotation, multiple cropping, habitat manipulation and use of trap crops are also effective practices which can keep the population of pests below economical threshold limit (ETL).

ii) **Mechanical alternative** - Removal of affected plants and plant parts, collection and destruction of egg masses and larvae, installation of bird perches, light traps, sticky colored plates and pheromone traps are most effective mechanical methods of pest control.

iii) **Biological alternative** - Use of pest predators and pathogens has also proved to be effective method of keeping pest problem. Inundative release of *Trichogramma* sp. @ 40,000 to 50,000 eggs per hectare, *Chelonus blackburni* @15,000 to 20, 000 per hectare, *Apanteles* sp.@15,000 to 20,000 per ha and *Chrysoperla* sp.@ 5,000 per ha., after 15 days of sowing and others parasites and predators after 30 days of sowing, can also effectively control pest problem in organic farming.

iv) **Biopesticides** - *Trichoderma viride* or *T. harazianum* or Pseudomonas fluorescence formulation @ 4gm/kg seed either alone or in combination can manage most of the seed borne and soil borne diseases. There are other formulations *viz.* Beauvaria bassiana, Metarizium anisopliae, Numeria rileyi, Verticillium sp, which are available in the market and can manage their specific host pest. *Bacillus thurengensis stenebrionis* and *B. thurengensis sandigo* are effective against coleopterans as well as some other insect species. Bt. has been used in the management of diamond back moth on crucifers and vegetables @ 0.5-1.0 kg. formulation per ha.

Viral biopesticides of baculovirus group *viz.* granulosis viruses and nuclear polyhedrosis viruses provide a great scope in plant protection field. Spray of nuclear polyhedrosis viruses of *Helicoverpa armigera* or *Spodoptera litura* @ 250 larval equivalents are very effective tools to manage the *Helicoverpa* sp. or *Spodoptera* sp. respectively.

v) **Botanical pesticides**: Many plants are known to have pesticidal properties and the extract of such plants or its refined forms can be used in the management of pests. Among various plants identified for the purpose, neem has been found to be most effective.

Neem (*Azadirachta indica*) – Neem has been found to be effective in the management of approximately 200 insects, pests and nematodes. Neem is very effective against grasshoppers, leaf hoppers, plant hoppers, aphids, jassids, and moth caterpillars. Neem extracts, are also very effective against beetle larvae, butterfly, moth and caterpillars such as Mexican bean beetle, Colorado potato beetle and diamondback moth.

Some other Pest Control Formulations

Many organic farmers and NGOs have developed large number of innovative formulations which are effectively used for control of various pests. Although none of these formulations have been subjected to scientific validation but their wide acceptance by farmers speak of their usefulness. Farmers can try these formulations, as they can be prepared on their own farm without the need of any purchases. Some of the popular formulations are listed below:

i) **Cow urine** – Cow urine diluted with water in ratio of 1: 20 and used as foliar spray is not only effective in the management of pathogens and insects, but also acts as effective growth promoter for the crop.

ii) **Fermented curd water** – In some parts of central India, fermented curd water (butter milk or Chaach) is also being used for the management of white fly, jassids aphids *etc.*

iii **Neem-Cow urine extract** – Fermentation of crushed neem leaves along-with cow dung, cow urine and water for 24 hrs make it useful against sucking pests and mealy bugs.

iv) **Mixed leaves extract** - Crushed neem leaves, custard apple leaves, papaya leaves, pomegranate leaves and cow urine are mixed in water. Boiling 5 times at some interval till it becomes half, make it useful against sucking pests, pod/fruit borers.

v) **Chilli-garlic extract** - Crushed Ipomea leaves, hot chilli, garlic and neem leaves in cow urine have also pesticidal activity against leaf roller, stem/fruit/pod borer.

Chapter 5

Ethics and Animal Welfare

The consumer's motivation for buying organic products is influenced by a concern for human health, environment and animal welfare. Animal welfare is often used but debated concept, both among philosophers and scientists. It refers to one or several aspects of an animal's quality of life, and a key question is how to interpret it, in theory as well as in practice. In organic farming, although the main concern is to develop sustainable and environmental friendly farming systems, animal welfare is an important aim. It is important for organic farmers and for the organic movement to take animal health and welfare issues seriously. Although most of the debate has been focusing on technical issues, one might suspect that the big differences among the antagonists lie in basic values and beliefs. That is, it may be a question of ethics.

Animal welfare can be seen from different perspectives or concepts; natural living, affective state or biological function and is multi-dimensional. Assessing animal welfare includes several welfare indicators that can be resource-based, related to the physical environment and management, or animal-based. The welfare indicators can be related to one or more of the welfare concepts. Animal-based welfare indicators measure the animals' response to the housing system, environment and management, which enables assessments of animal welfare across different systems.

Animal welfare can be defined in different ways. Broom (1996) defined animal welfare as the animal's attempt to cope with the environment, and Duncan (1996) stated that animal welfare depends on the animal's feelings. These views are also found in the three welfare concepts identified by Fraser *et al.* (1997):

i) **Natural living**: Animals should freely be able to carry out normal behavior, be free from constraints and live a natural life, *e.g.* outdoor access and having an intact beak

ii) **Affective state (feeling and emotions)**: Pain and suffering should be avoided and positive states should be emphasized

iii) **Biological function:** Productivity and health of the animal are important

Broom (1996) and Duncan's (1996) definitions are in line with the concept of biological function and affective states, respectively. Another view of animal welfare is the five freedoms (FAWC, 1979).

i) Freedom from thirst, hunger and malnutrition

ii) Freedom from discomfort

iii) Freedom from pain, injury and diseases

iv) Freedom to express normal behavior

v) Freedom from fear and distress

The three welfare concepts and the five freedoms have similarities and, to some extent, cover the same aspects. Freedom from thirst, hunger and malnutrition is included in the concept of biological function, freedom from discomfort is included in the concept of affective state, freedom from pain, injury and diseases is included in the concept of biological function and affective state, freedom to express normal behavior is included in the concept of natural living and affective state, and freedom from fear and distress is included in the concept of affective state.

Delivering Enhanced Animal Welfare Benefits

Animals in agricultural systems to a large extent depend on the human caretaker for their existence and living conditions. Conventional production is typically aiming for high production levels with restricted input resources such as space, feed *etc.*, and these conditions may cause stress in the individual animal as it is unable to cope with the situation, *e.g.* in pig production. This means that higher stocking density, restricted space and barren environment are factors increasing the risk of the development of diseases, and therefore it is more likely that animals under these conditions need antibiotic treatments.

Organic production aims for less intensive animal production, which generally means that the animals have access to a more spacious and enriched environment, access to an outdoor range and restricted group sizes, and other preconditions. This would ultimately decrease the need for preventive medication of the animals as they can perform more natural behaviors and have more opportunity to maintain a good health. However, in practice, the health status of organic livestock is complex and disease prevention needs to be adapted to the individual farm.

Under the rules governing organic production, higher standards of animal husbandry and associated animal welfare benefits are prescribed in the codified organic regulations and standards. These are as follows:

- Housed animals must be provided with bedding.
- Good ventilation and a generous floor area for each animal are required.

- Livestock must be fed with organic feed that meets the animal's nutritional requirements at the various stages of development and must have constant access to permanent pasture and roughage.

In organic systems, animal health is seen not simply as the absence of disease; it is seen as a positive characteristic which is to be achieved through the application of biological and animal husbandry principles rather than the routine use of conventional veterinary medicines. While the emphasis is on disease prevention through enhanced animal welfare measures, routine preventative treatment of healthy animals is not allowed. Disease prevention is based on breed and strain selection, husbandry management practices, high quality feed and exercise for optimal natural resistance to combat disease, appropriate stocking density and adequate and appropriate housing maintained in hygienic conditions. The standards for organic livestock production emphasize preventative strategies based on the principles that an animal is allowed to exhibit natural behavior and is not subject to stress. Where medicines are required, the use of complementary medicines both for the prevention and treatment of disease is encouraged. Failure to treat sick animals may result in the withdrawal of organic status for the entire farm. When a farm undergoes conversion to organic status, an Animal Health Plan is required to be drawn up by the veterinary practitioner. This specifies the current animal health issues on the farm and how the farmer will tackle these problems into the future while conforming to the requirements of organic certification standards.

Chapter 6

Organic Livestock Farming

Animal production is an important part of organic farming that aims at achieving a balanced relationship between the soil, the plants and the animals in a farming system. Consumers expect the food from organic production to be of a certain quality that makes it different from conventionally produced food. Acknowledged consumer interests cover a wide range of issues, from the nature of farming as a whole (environmentally friendly, socially just, animal welfare friendly) to the concern over own health (buying organic food based on a perception of organic food being more healthy). Each component is as important as the other in contributing to the overall effect and in fulfilling the key values of naturalness, harmony, local circulation of resources and the principle of precaution.

Organic livestock production in general is a land based activity and shall be an integral part of organic farm unit and management of livestock shall be in consistent with the principles of organic farming and shall base on:

a) Natural breeding
b) Protection of animal health and welfare
c) Fed with organic feed and fodder
d) Access to grazing in organic fields
e) Freedom to express natural behaviour
f) Reduction of stress
g) Prohibition of use of chemically synthesized allopathic veterinary drugs, antibiotics, hormones, growth boosters, feed additives *etc.*

In an organic system, the welfare of the animals is considered very important. Animals raised using organic methods are treated humanely and with respect. In order to achieve the animal welfare, environmental, resource-use sustainability and other objectives, certain key principles are as follows:

- ☆ Management of livestock as land-based systems (*i.e.* excluding feedlots and intensively-housed pig and poultry units) so that stock numbers are related to the carrying capacity of the land and not inflated by reliance on 'purchased' hectares from outside the farm system, thus avoiding the potential for nutrient concentration, excess manure production and pollution.
- ☆ Reliance on farm- or locally-derived renewable resources, such as biologically-fixed atmospheric nitrogen and home-grown livestock feeds, thereby reducing the need for non-renewable resources as direct inputs or for transport.
- ☆ Reliance on feed sources produced organically, which are suited to the animal's evolutionary adaptations (including restrictions on use of animal proteins) and which minimize competition for food suitable for human consumption.
- ☆ Maintenance of health through preventive management and good husbandry in preference to preventive treatment, thereby reducing the potential for the development of resistance to therapeutic medicines as well as contamination of workers, food products and the environment.
- ☆ Use of housing systems which allow natural behavior patterns to be followed and which give high priority to animal welfare considerations, with the emphasis on free-range systems for poultry.
- ☆ Use of breeds and rearing systems suited to the production systems employed, in terms of disease resistance, productivity, hardiness, and suitability for ranging.

Health and Medication

The aim of organic systems is to optimize breeding, rearing, feeding, housing and general management in order to achieve stability and balance in the farming system, maximize the health of the animal and minimize disease pressure and stress. In organic standards preventive treatment is restricted to the restrained use of vaccination and homoeopathic nosodes for known farm problems. Growth promoters, hormones and the routine (prophylactic) use of antibiotics are not allowed. All organic standards emphasize the prevention of disease by enhancing the welfare of the animal and prohibit the use of prophylactic treatment with chemotherapy. Where possible, treatment of ailments is approached by aiding the animal's own resistance and the use of complementary therapies such as homoeopathy. Conventional treatment should be used in all cases where it is necessary to prevent prolonged illness or suffering. The draft EU organic livestock regulation allows a maximum of two courses of treatment in an annual production cycle, or lifetime if that is less than one year, above this they will lose their organic status for that year.

Choice of Breeds, Source and Origin

Choice of Breeds

The choice of livestock and poultry, breeds, strains and breeding methods shall be consistent with the principles of organic farming, taking into account, in particular, the following:

i) Their adaptation to the local climatic conditions and

ii) Their vitality and resistance to diseases

Sources/Origin

i) Transfer of livestock and poultry between organic and non-organic units shall not be permitted.

ii) Livestock and poultry raised on non-organic production units shall be converted to organic.

iii) When a producer demonstrates to the satisfaction of the accredited certification body that the organic source livestock are not available, the accredited Certification Body may allow such livestock and poultry under the following circumstances:

   a) When the producer is establishing an organic livestock and poultry operation for the first time.

   b) When the producer wants to change the livestock and poultry breed/ strain or when new livestock and poultry specialization is developed.

   c) For the renewal of a herd, *e.g.*, due to high mortality of animals caused by catastrophic circumstances.

   d) When the producer wishes to introduce breeding males into the farm.

In all such cases product of such animals shall qualify for organic only after completion of conversion period.

Livestock Identification and Animal Record Keeping

Livestock Identification

i) Each animal/herd/batch shall bear unique identification number. Large animals including bovine, ovine, carpine, porcine *etc.* shall bear individual number in the form of tag, while poultry birds and small mammals shall be identified with herd/flock/batch.

ii) Identification devices on the animals can be printed ear tags, RFID tags, barcodes or any other suitable tag which is clearly visible.

Record Keeping

Following data for each animal/herd or batch shall be maintained and made available to the accredited certification body for verification during inspection:

i) Parent details

ii) Source and purchase details

iii) Animal details

iv) Breeding details

v) Feeding details

vi) Health care details including details of vaccination, medication, veterinarian prescription and withdrawal period *etc.*

vii) Production details

viii) Sale details

ix) Any other relevant details.

Housing and Management

i) The housing and day-to-day management of the animal, maintenance of sanitation, hygiene, bio-security and environment shall be planned to suit the specific behavioral needs of the livestock and poultry and shall provide for sufficient space to ensure free movement and opportunity to express normal patterns of behavior.

ii) The animals should not be tied; however animals can be confined for specific reasons, such as, milking, for some medical procedures, controlled grazing, during night time and for health and safety of animal.

iii) Where the livestock and poultry's normal behavior demands group living, animals shall not be kept in isolation, but shall have company of like kind.

iv) As far as possible two different kinds of animals shall not be kept together, unless for specific purposes, such as, free range poultry birds in cow/ buffalo shed for scavenging on ticks and other insects.

v) The housing system shall ensure prevention of abnormal behavior, injury and disease.

vi) Appropriate facilities to cover emergencies such as the fire, the breakdown of essential mechanical services and the disruption of supplies, shall be available.

vii) Housing for Livestock and Poultry shall not be mandatory in areas where appropriate climatic conditions exist to enable animals to live outdoors without compromising their comfort, health and welfare. Conditions shall be inspected and permitted by the accredited Certification Body on producer and location- to- location basis. Outdoor open areas may be partially covered.

viii) Housing conditions shall meet the biological and behavioral needs of the livestock and poultry by providing easy access to feeding and watering and shall ensure:

a) Insulation, heating, cooling and ventilation of the building to ensure that air circulation, dust level, temperature, relative air humidity and gas concentrations are kept within limits which are not harmful to the livestock and poultry.

b) Plentiful natural ventilation and light to enter.

c) Appropriate fencing not harmful to the animals.

ix) Confinement shall be permitted under following conditions:

a) Inclement weather to protect animals from injury.

b) Ensure health safety or welfare.

c) To protect plant, soil and water quality.

x) The outdoor stocking density of livestock kept on pasture, grassland, or other natural or semi-natural habitats, must be low enough to prevent degradation of the soil and over-grazing of vegetation.

Special Conditions for Mammals

i) All mammals shall have access to open-air exercise or resting area, paddock, pen or run which may be partially covered and/or shall have space for protection from rains and hot sun.

ii) Livestock shed shall have properly laid and smooth floor, although not slippery. The floor shall not be entirely of slatted or grid construction.

iii) The housing conditions shall aim at providing comfortable, clean and dry laying/rest area of sufficient size, consisting of a solid construction. Wherever possible, straw bedding shall be provided.

iv) The calves may be housed separately and never in the adult animal shed.

v) Pigs must be kept in groups, except in the last stages of pregnancy and during the suckling period. Piglets may not be kept on flat decks or in piglet cages. Exercise areas must permit dunging and rooting by the animal. Breeding boars may be kept separately.

Special Conditions for Rabbits

i) The keeping of rabbits in cages shall not be permitted. If required for comfort and safety rabbits may be temporarily confined, for example overnight, in cages or hutches. Continuous confinement is prohibited.

ii) Rabbits shall have space to run, hop and dig, and to sit upright on their back legs with ears erect.

Conversion Period for Animal Production

i) When a livestock production unit, with entire herd, or flock of sheep/ goat or batch of poultry birds or small mammals such as rabbits, is in transition to organic production, pasture and feed produced on the land undergone a minimum period of 12 months of conversion period may be considered organic for feeding to organic livestock.

iii) The conversion period shall be determined by the accredited Certification Body and the conversion period shall be accounted from the date of first inspection.

iv) In cases, where the land and livestock and poultry conversion to organic status is not simultaneous and the land alone has reached organic status and the livestock and poultry from a non-organic source is introduced, these must be reared according to the guidelines for at least the following compliance periods before their products are sold as organic.

a) Bovine including Buffalo

i) Meat products: Twelve months and at least 3/4th of their life span is spent in the organic management system.

ii) Calves for meat production: Six months when brought in as soon as they are weaned and less than six months old.

iii) Milk products: Six months.

b) Ovine and Caprine (Sheep and Goat)

i) Meat products: Six months.

ii) Milk products: Six months.

c) Pig

i) Meat products: Six (6) months.

d) Small Mammals (Such as Rabbits)

i) Meat products: From the second week after their birth to the entire life span as determined by the accredited Certification Body.

e) Poultry

i) Meat products: from the second day of hatching to the entire life span as determined by the accredited Certification Body.

ii) Eggs : Six weeks.

Feed

Livestock and poultry farms shall provide maximum diet from feedstuffs (including 'in conversion' feedstuff) produced as organic. Agricultural processed residues of organic origin, such as from grain fermentation, fruit processing, vegetable processing, *etc.*, shall be permitted for purpose of feeding, provided that the overall feeding practices satisfy the daily energy and nutrient requirements of the concerned animals. The agriculture land committed to cultivation of feed/fodder crops intended to be used as feed for livestock and poultry shall be organically grown.

During the operations, the products shall maintain their organic status provided that livestock and poultry are fed with at least 85 per cent for ruminants and 80 per cent for non-ruminants calculated on a dry matter basis, feed obtained from organic sources.

Specific Livestock and Poultry Rations shall take into Account

i) The need of young animals for natural feed, such as, feeding of maternal milk, milk from other mammal or milk replacer of organic origin that has maximum similarity with maternal milk, provided that it does not contain any genetically modified ingredient, antibiotics, hormone, *etc.*

ii) That in herbivores, substantial proportion of the dry matter and energy in the daily rations should consist of roughage, fresh or dried fodder, or silage; need for inclusion of cereals in the fattening phase of poultry and livestock and poultry must have ample, free access to water appropriate to maintain full health and productivity.

Specific Criteria for Additives and Processing Aids

i) The supplements should be derived from natural sources.

ii) Feed processing aid supplements like binders, anti-caking agents, emulsifiers, stabilizers, thickeners, surfactants, coagulants if used should be from natural sources.

iii) Antioxidants: only from natural sources shall be permitted.

iv) Preservatives: only natural acids are allowed.

v) Colouring agents (including pigments), flavors, odor masking agents and appetite stimulants: only natural sources are allowed.

vi) Probiotics, enzymes and microorganisms are allowed but should not be from genetically modified sources.

vii) Any synthetic chemicals, such as, antibiotics, coccidiostat, medicine, growth promoters or any other substance supplemented for purpose to stimulate growth or production shall not be fed to the organic livestock and poultry.

viii) Silage additives, additives for enriching crop residues and processing aids may not be derived from genetically engineered/modified organisms or products thereof, and may be comprised of only:

   a) Sea salt

   b) Coarse rock salt

   c) Yeasts

   d) Enzymes

   e) Whey

   f) Sugar; or sugar products such as molasses, jaggery, grain bran

   g) Honey

   h) Lactic, acetic, formic and propionic bacteria, or their natural acid product when the weather conditions do not allow for adequate fermentation and their use to be approved by the accredited Certification Body.

Health Care

The organic livestock in general, should follow the basic principles of preventive health and productivity management wherein the focus would be on preventing diseases, detecting underlying fertility and production problems and its correction primarily on correcting management, nutrition and sanitation. The use of veterinary medicinal products in organic farming shall comply with the following principles:

a) All vaccinations required by law of the land shall be permitted. Where specific disease or health problems occur, or is predicted to occur, and there are no alternative permitted treatment or management practice exist, use of parasiticides, or therapeutic use of veterinary drugs are permitted under prescription and supervision of a registered veterinarian.

b) For purpose of treatment and prevention of diseases and under performances, herbal/phyto-therapeutic (excluding antibiotics), homeopathic or ayurvedic products shall be preferred to allopathic veterinary drugs or antibiotics, provided that their therapeutic effect is effective for the species of animal and the condition for which the treatment is intended.

c) In case alternative therapeutic or preventive measures are unlikely to be effective in combating illness or injury, allopathic veterinary drugs or antibiotics may be used under the responsibility and supervision of a veterinarian.

d) The use of allopathic veterinary drugs or antibiotics or drugs derived from genetically modified source for preventative treatments and for enhancing productivity or fertility is prohibited.

e) Hormonal treatment may only be used for therapeutic reasons and under veterinary supervision.

f) Growth stimulants, agents or substances used for the purpose of stimulating growth or production shall not be permitted.

Breeding

a) The breeds and strains most suited to local conditions.

b) The preference for reproduction through natural methods, although artificial insemination may be used.

c) Embryo transfer techniques and any other breeding techniques employing genetic engineering shall not be used.

d) The use of hormonal reproductive treatment shall not be used unless prescribed therapeutic, directed towards correcting the physiological problem.

Management

Mutilation, such as, tail docking, cutting of teeth and dehorning are not permitted. In exceptional cases, some of these may be authorized by the accredited

Certification Body for reasons of safety (*e.g.* dehorning in young animals, hoof trimming, cutting of pin teeth in pigs *etc.*) or if they are intended to improve the health and welfare of the livestock. Such surgical procedures shall be carried out by a registered veterinarian at the most appropriate age; and any suffering to and pain shall be reduced to a minimum. Wherever possible, anesthetic and analgesics shall be used. Physical castration is allowed only in order to maintain the quality of products and traditional production practices (meat-type pigs, bullocks, capons, *etc.*).

Organic Animal Feed Processing and Handling

Organic animal feed processing and handling includes processing of organic feed and food for all types of domesticated animals including livestock, poultry, aquaculture and pet animals for production of commercial animal feed/food products. However, use of following is prohibited in organic products:

- ☆ Genetically modified organisms or ingredients originating from genetically modified organisms are prohibited
- ☆ Synthetic chemicals used for boosting metabolism is prohibited
- ☆ Synthetic nitrogen or non-protein nitrogen compounds are prohibited
- ☆ Antibiotics, synthetic antimicrobials, growth enhancing substances, parasiticides, coccidiostatics or hormones
- ☆ Other substances produced or modified through artificial synthesis
- ☆ Feed or raw material of mammalian origin including slaughter house waste for making feed for ruminant livestock
- ☆ Irradiation for food processing is not allowed
- ☆ Packaging methods and materials must protect the integrity of organic feed and should not have any adverse effects on the environment.
- ☆ Packaging materials, containers and storage containing or treated with synthetic chemicals or prohibited substances must not be used.

Processed Products

100 per cent organic - A raw or processed animal feed sold, labeled, or represented as "100 per cent organic" must contain (by weight or fluid volume, excluding water and salt) 100 per cent organically produced ingredients.

Organic - A raw or processed animal feed sold, labeled, or represented as "organic" must contain (by weight or fluid volume, excluding water and salt) not less than 95 per cent organically produced raw or processed agricultural ingredients.

Made with organic - Multi-ingredient animal feed sold, labeled, or represented as "made with organic (specified ingredients or food group(s))" must contain (by weight or fluid volume, excluding water and salt) at least 70 per cent organically produced ingredients.

Where less than 70 per cent of the ingredients are of certified organic origin, the indication that an ingredient is organic may appear in the ingredients list. Such product may not be called "organic.

Pest Control in Organic Farming

Pests should be avoided by good manufacturing practices. This includes general cleanliness and hygiene. Treatments with pest regulating agents must thus be regarded as the last resort. Recommended treatments are physical barriers, sound, ultra-sound, light and UV-light, traps (*e.g.*, pheromone traps and static bait traps), temperature control, controlled atmosphere and diatomaceous earth. Irradiation is prohibited. Persistent or carcinogenic pesticides and disinfectants are also not permitted.

Chapter 7

Organic Poultry Farming

Poultry Production Methods

In recent years, several different methods or systems have evolved for producing eggs. These systems vary in how the birds are housed, fed and managed. Before discussing organic poultry production it is important to clearly identify the various systems.

Conventional Layers (Battery Cages)

In this system, two to four pullets as young as 19 weeks of age are put into a wire cage with an area of approximately two square feet. A barn full of these cages is known as a battery of cages. These barns contain thousands of birds. Feeds are allowed to contain antibiotics at prescribed levels.

Free Run

This system is similar to conventional layers but does not use battery cages. The layers are allowed to run on the floor of a barn. Birds are densely packed in these barns with no access to the outdoors, and are fed the same feed as battery birds. Eggs produced by this system would be known as 'free run' eggs.

Free Range

Birds in this system are required to have outside access most of the year. Birds would have a significant outside run or runs, and would have roosts for resting. There is the same floor space requirement as organic. The feed is not allowed to contain any antibiotics or production stimulants. Eggs produced in this system would be known as 'free range' eggs.

Pastured

The layers are kept in a movable enclosure with nests and roosts. The structure is moved once or twice daily to a new piece of grass. The chickens will forage and get 20 per cent of their diet from the forage and insects in the new ground daily.

This system differs from free range as the birds have access to new grass/ground daily. In this region birds can be pastured for six to seven months a year.

Organic

Organic guidelines are more stringent than free range. The pullets must be raised by certified organic production methods from birth. The layers are required to have outdoor access year round, or be fed sprouted grains for the period when indoors only. One can buy conventional chicks but they have to be raised organically from birth to be considered organic birds.

Production Considerations

Organic Chicks

Chicks have traditionally been raised on finely ground boiled eggs mixed with oatmeal. This type of starter can be made on farm provided the eggs and oatmeal are certified. Also, any poultry diet can be improved significantly by the addition of cod liver oil. This may be too expensive. The birds are out of production for two months during the molting process and produce well for another ten months following the molt. These birds can be quite productive for two years and molting reduces input costs.

In general, organic production standards are more comprehensive and restrictive than most 'intermediate' standards. Purchasing from commercial hatcheries means that eggs and chicks will have undergone precautionary hygiene treatments needed in large-scale hatcheries. This may include formaldehyde treatments of eggs, and if chicks hatch soon after treatment there may be some implications for later respiratory problems.

Typically, pullets will have received vaccinations for Newcastle disease (fowl pest), infectious bronchitis and Marek's disease. Some rearers will also vaccinate against ILT (infectious laryngotracheitis) at 11-16 weeks of age. ILT can be spread by vaccine, so in any case producers should ensure that they do not buy vaccinated birds if they do not have the infection on their unit.

Organic Poultry

Organic poultry must have daily access to the outdoors, as weather permits. Access areas are not mandated to have vegetation providing feed value, as with ruminants. However, access to pasture can be a selling point and is claimed by some to produce healthier birds and better-tasting poultry products. Any land the birds have access to must be certified organic (no prohibited materials for three years).

Table Birds

As table birds have to be purchased as day-old chicks, the young chicks need to be kept under brooders for the first 4-5 weeks in purpose-built housing. Gas brooders are preferred to electric red light heaters by producers on grounds of reliability and flexibility due to the risk of power cuts and bulbs blowing, and increased mobility. Some producers have also found smothering to be less of a problem with gas brooders. The main requirement to reduce smothering is a diffuse heat source to avoid huddling. Rounding off corners using circular partitions also helps to avoid huddling in corners. Accommodation should be well insulated, with uniformity of light intensity, heat and ventilation to encourage use of the whole floor area.

Management Practices in Organic Poultry Farming

Housing

Housing must allow exercise, freedom of movement and reduction of stress. Hens must be provided with natural light. Stationary houses, moveable pens or moveable houses may be used. Layers may not be caged and must have access to the ground or floor. Some buyers require housing that allows at least 1.75 square foot of floor space per bird inside the hen house, and 2 square foot per bird outside. Floor should be covered with clean straw, wood shavings, or sawdust. Hens enjoy roosting at night, which may be made of wood, plastic or metal. If roosts are used, at least 50 per cent of the henhouse should allow access to the floor to allow scratching.

One feed pan usually provides enough space for 20 birds. Sufficient watering equipment should be available for 20 birds per cup, 12 birds per nipple, or 1 bird per linear inch of trough space. Birds do not lay eggs at the same time, so nests can be supplied at the rate of one nest per five hens. Nest box bedding should be different from floor litter and must be kept clean and dry.

Ventilation of poultry enclosures is necessary so that adequate air exchange can take place to keep littered floors dry. Ventilation needs will vary with ambient temperature. Misting water around the outside of poultry enclosures may be necessary to keep layers cool during the extreme heat of summer months. Additional litter materials may be necessary to provide insulation during the winter to protect hens from cold floors and drafts.

If bedding is used during brooding or housing, it must be certified organic, if it is agricultural and an item the poultry will typically consume (*i.e.* hay). Typical bedding may be wood shavings (not from treated wood), organic corncobs, organic hay or straw or organic corn fodder.

Organic standards aim to provide an environment for poultry in which all normal behavior patterns can occur as this will minimize the stress to the birds. Low stress levels are likely to have a positive effect upon both the health and production capacity of the flock. The majority of producers shut their poultry in at night to protect them from predators and therefore the design of the housing system must take the behavioral needs that arise during this period of confinement into account. Two approaches to housing are generally used by organic producers: mobile systems with houses that can be moved, for example to utilize the grass/clover ley in an arable rotation, and static housing systems, where the birds have access to an outside area covered with vegetation.

Attractive nests reduce the number of ground and mislaid eggs and therefore the number of dirty eggs and the time needed to collect the eggs. Nests should be in a quiet corner of the barn, preferably away from light and pop-holes. In semi-intensive systems with up to 12 hens/m^2 perches are situated above a mesh covered area where droppings are collected. For table birds, the use of perchery housing is less appropriate, because of the problems with weight relative to leg strength. This is also a potential problem with the use of outside feeders referred to above. Reducing the contact of the birds with their droppings by covering the litter area reduces the risk of infection. In more intensive systems, the droppings are automatically collected, usually by conveyor belts under the perches. Regular removal of droppings also helps to improve air quality in larger houses. About 25-33 per cent of the floor area should be available for scratching. The litter material can consist of straw, soft wood shavings, compost and sand, or a combination of these. It should be kept dry and friable to minimize infection, but not too dry to avoid excessive dust in the air. Humidity should be kept below 70 per cent. The layer of bedding should not be too thick so that the claws can wear off. Sand and dust bathing facilities can be offered additionally if not available in the outside area. The Farm Animal Welfare Council (FAWC) recommends one meter of nesting space per 80 birds, but may be reduced to 1 meter per 120 birds. Feeding and drinking equipment needs to be situated where it is not contaminated by excrement. In larger systems automatic feeding and drinking equipment and also automatic collection of eggs is useful to limit the time needed otherwise for those activities.

The following points are needed to be pursued for organic poultry farming:

- Hens are kept in floor or aviary systems in flocks with at maximum 3000 hens per flock.
- The hens have access to an outdoor area, an indoor area with a littered floor area, perches, nests and roughage. These production conditions promote natural behavior and potentially increase animal welfare.
- The hens should be housed with an indoor stocking density of maximum 6 hens/m^2 with one third of the indoor being a solid littered area, have access to a free-range.
- The outdoor area should always count minimum 4 m^2 per hen per rotation, and the egg producers can fulfill the requirements in four possible ways:
    - 6 m^2 per hen. In this case paddock rotations are not necessary. However, the free-range area should be covered with trees and bushes to give a forest or plantation-like appearance
    - Gradual paddock rotation, where the hens always have access to 4 m^2 per hen
    - 4 m^2 per hen, where the free-range area has a 60-day period during the growing season every year without hens
    - 4 m^2 per hen, where the free-range area has a 120-day period during the growing season every second year without hens

Lighting

Lighting stimulates hens to lay eggs. To produce eggs year-round, adequate lighting should be installed. Gradually the length of time hens are exposed to light should be increased when they arrive at farm. 12 hours of total light per day at an indoor intensity (0.5- to 1.0-foot candles) should be there. Daylight length should be increase by 30 minutes per week until it is reached 16 hours of light per day. Additional outside light exposure is fine; one should just have the supplementary lights on before and after dawn and dusk.

Organic standards specify that natural lighting must be available. However, natural light (direct sunlight in particular) is considered to increase the potential for feather pecking and cannibalism. Careful design of the windows is therefore required, so that equal distribution of the daylight can be achieved. For this purpose the use of windows in the roof can be beneficial. However, because of the above mentioned risk direct sunlight should be avoided, either by placing extended roofs in front of the windows or by covering the glass with white paint. Artificial lighting can be used to extend the lighting period to a maximum period of 16 hours of light, whereby the day length should be extended in the morning rather than in the evening. In this way the majority of hens will have laid before they go out, so that the chances of dirt contamination of the eggs are reduced.

Fencing

It is not strictly necessary to have fencing around the range areas as birds will not move far from their houses and they are normally shut up at night. However, if predators are a problem or there is a need to protect other crops this should be considered. The most effective form of fencing is electric netting connected either to batteries or to the mains, although the losses to predators might be greater than with a fixed fence. Stands of electrified wire at 10 cm intervals can be used as an alternative to netting. If a permanent non-electrified fence is used, it should be 2 m high, with a portion buried horizontally to the outside and an electric wire 30 cm away from the fence and above the ground to deter digging by foxes.

Feed

Feed must be either purchased as certified organic or produced on own certified organic farm. Any agricultural products in feed or feed supplements must be organic. This includes secondary ingredients such as soy oil or wheat middlings. Non-agricultural natural ingredients such as kelp, grit, calcium or fishmeal must be approved for use in organic operations. No synthetic preservatives, colors, flowing agents or dust suppressants are allowed. Feed may not include mammalian or poultry slaughter byproducts. Adding flax meal to ration can increase the presence of omega 3 fatty acids in the eggs.

Organic production standards specify that only a limited range of purchased feedstuffs may be used, that 70-80 per cent of the ration should be certified organic or in conversion, and that ideally poultry should have access to vegetation (green fodder). Ideally, a high proportion of the feedstuffs supplied should be grown on the farm itself. This reduces the pollution problems associated with imports of large quantities of nutrients and their disposal as livestock wastes on relatively small areas of land. If all feedstuffs would be grown on the farm, one hectare of farmland could support approximately 100 layers. These requirements have implications for ration formulation to meet production targets, in particular protein balance, as well as cost since organic cereals attract a significant premium. Additives such as the antibiotic Avoparcin, are prohibited due to fears of transmission of antibiotic resistance to bacteria infecting humans, are not permitted. A protein concentrate with mineral and vitamin supplements can be fed separately. Calcium sources (soluble grit) should be fed separately as a third food choice to avoid overeating of protein at times of high calcium requirement.

Conventional layer mash feed can be purchased at your local feed store. Certified organic feeds are also available, but they are sometimes more difficult to find and are more costly. All hen mash should contain at least 3.5 per cent calcium and 16 per cent crude protein. Additional free-choice calcium may be provided after birds are 45 weeks old to aid in good shell formation, especially during hot weather. Water should be given free choice, and container-based watering systems should be emptied and cleaned every other day at a minimum. A bucket should be used to carry waste water away from living areas to reduce the potential for disease and to prevent the litter from becoming damp or wet.

IFOAM is intending to revise the standards to prohibit the use of synthetic amino acids in organic rations. Corn-fed table birds may be fed a ration containing a minimum of 65 per cent cereals, a maximum of 15 per cent cereal by-products, a maximum of 5 per cent pulses or green vegetables and a maximum of 5 per cent dairy products. If specific cereals are named, they must comprise at least 35 per cent of the ration, 50 per cent in the case of maize. The use of growth promoters and coccidiostats, essentially following free-range and in some cases organic standards is prohibited, but without the requirement for organic feed or the costs of independent certification.

All organic standards state that ideally 100 per cent of the diet should be organic; however, some allowance for some non-organic components of the diet may be given. At least 50 per cent of the diet must be fully organic, and the remaining 20 per cent can come from sources that are in conversion to organic production.

Searching for food and feeding have an important social function for the hens, and the presence of other feeding hens stimulates feeding behavior. Additionally structure and color of the food encourage (or discourage) the hens to feed. Larger particles are more easily recognized than fine dust, bright colors are preferred. Hens have a muscular stomach but no teeth. Eating of sand or small stones helps the digestive system. Feeding some whole grain as scratch feed in the litter or outside areas allow the hens to exercise their natural behavior, but should be restricted so that the hens still consume enough feed with higher protein and energy content. If whole grain is fed, access to fine stones (*e.g.* insoluble flint or granite grit, not limestone) should be given. Soaking some of the grain prior to feeding increases food intake (as the hens show a preference for moistened grain) and to increase the time occupied by feeding.

One organic producer uses a starter ration with 20-22 per cent crude protein for the first 4 weeks for table birds, followed by a finishing ration with 18 per cent protein to finish at 10-12 weeks, giving an overall feed conversion factor of 3.5-4. In the last couple of weeks, some wheat is added to the rations to get a thin layer of fat to improve flavor.

Medicated Feeds including Hormones or Antibiotics are not Allowed in Organics

Health issues should be treated through prevention, as most poultry diseases are very difficult to treat. Sanitation is the best form of defense in disease management. Healthcare alternatives include vinegar in water or turmeric added to food for coccidiosis, homeopathy, probiotics, herbs or hydrogen peroxide. Agricultural substances added daily/routinely to feed or water such as vinegar, turmeric, *etc.* (not just to treat medical situations) must be certified organic.

Farm Biosecurity

It is very important for preventing transfer of diseases from farm to farm, from wild birds to domestic birds and from one batch of poultry to another. Biosecurity and sanitation are necessary to prevent disease outbreaks. Biosecurity involves isolating birds by age-group, restricting human access to buildings, keeping the buildings clean, and properly disposing of dead birds. To prevent the introduction of diseases, new birds should be isolated and observed for disease symptoms for one month before allowing contact with other birds.

Because of cannibalistic tendencies often seen in birds, housing of flocks should be done in an all-in, all-out fashion so that birds kept together are of the same age and size. If smaller birds are being raised as replacements, they should be kept in separate housing and seen first each day. Dead birds should be composted in a compost pile large enough to cover the birds with 6 inches of composting material surrounding the body or double-bagged if disposed of in municipal waste receptacles.

Consulting a veterinarian regularly is a good management practice. If high mortality suddenly appears in flock with no apparent cause, veterinarian should be informed at state department immediately.

Vaccines

Vaccines are allowed, although are mostly used in larger operations. Typical vaccines that may be considered include vaccines for: *Newcastle disease, coccidiosis, mycoplasma, gallisepticum* and *Mycoplasma synoviae*. Vaccines may be administered via water, through the air or orally.

Parasite Control

Laying hens may show symptoms of lice predation. Lice can be controlled though cleanliness in the hen house, including the use of bleach or lime in nest boxes. Providing dust baths of diatomaceous earth are used by some for lice control.

Cannibalism

Cannibalism may be caused by overcrowding or ration imbalance. These conditions may be corrected to reduce the problem. Poultry will peck at bloody spots, so if a bird is injured it should be isolated from others.

Predators

Predators are often a significant factor for small-scale organic poultry production. Predators can include raccoons, dogs fox, coyotes, mink, weasels, opossums, rats and aerial predators including owls and hawks. Poisons are not allowed. Common control methods include tight structures, electric fencing, guard animals and flashing lights.

To Maintain an Economically Viable Laying Flock

Management must take into consideration careful timing of the introduction of young pullets for consistent egg supply.

Egg Collection

Eggs may be collected from traditional, hand-gathered nest boxes or from roll-away nests. Eggs may be fertilized (rooster present) or not. Eggs may be cleaned with approved cleaners, using water that is at least 10 degrees warmer than the eggs, as long as the residue of the cleaner left on the eggs has been rinsed off with potable water. Re-use of egg cartons is allowed as long as no residue of prohibited materials. Dirty, leaker, or loss eggs should be removed. Loss eggs are inedible or contaminated eggs discovered when held to a bright light (candling). There are three consumer grades of eggs—Grade AA, Grade A, and Grade B. To market eggs on these terms, they must meet the requirements for the consumer grade.

Health Care Management

The aim of organic systems is to optimize breeding, rearing, feeding, housing and general management in order to achieve stability and balance in the farming system, maximize health of the animal and minimize disease and stress. In organic standards, preventive treatment is restricted to the restrained use of vaccination and homoeopathic medicines for known farm problems. Growth promoters, hormones and the routine (prophylactic) use of antibiotics are not allowed. All organic standards emphasize the prevention of disease by enhancing the welfare of the animal and prohibit the use of prophylactic treatment with chemotherapy. Where possible, treatment of ailments is approached by aiding the animal's own resistance and use of complementary therapies such as homoeopathy. Conventional treatment should be used in all cases where it is necessary to prevent prolonged illness or suffering, but longer withdrawal periods are imposed under organic production standards for controlled drugs and veterinary products. Beak trimming is not allowed.

Documentation

Records must be kept on source and age of poultry, feed and supplement use and sources, use and source of any health products, vaccinations, mortalities, outside access, house sanitation practices between flocks, collection of eggs and sale of finished products.

IFOAM Standards for Organic Production for Poultry

Pop holes	Continuous and easy access to runs
General welfare	Behavioral needs must be provided
Beak trimming	Not permitted
Wing clipping	Not permitted
Feeding	Must be produced on the farm or from organic farms in the region (exemption possible)
Feed origin	Max 20 per cent conventional
Animal protein	Local certifier to specify
Pure amino acids	Local certifier to specify
Growth promoters	No
Health	Maximize resistance and prevention
Medications	Double withdrawal period
Prohibited	Prophylactic treatment, growth promoters
Transport	Must not result in physical injury, no chemically synthesized tranquilizers
Journey time	Maximum 8 hrs to slaughter house
Age of chicks for stock origin (table birds)	1 or 2 days
Housing	To allow behavioral needs
Littered area (layers)	Natural materials required
Littered area (table birds)	Natural materials required

Animal welfare is multi-dimensional, and a welfare assessment should therefore include several indicators to give a comprehensive view of the current situation. The consumers' motivation to buy organic products includes animal welfare aspects; there are a number of welfare issues in the organic egg production compared to other production systems, like higher mortality and prevalence of helminth infections.

The welfare of the hens according to a set of criteria that must be fulfilled to have the welfare evaluated as satisfactory:

- ✰ The mortality must not exceed 1.2 per cent per month
- ✰ The proportion of hens that is severely feather pecked should not exceed 5 per cent (defined as a hen with two or more nude areas of δ5 cm in diameter)
- ✰ The proportion of hens that is moderately feather pecked changes with age. At 25 weeks of age, the proportion of moderately feather pecked hens should not exceed 10 per cent, and this level increases with 1 per cent per live week, *i.e.* at 65 weeks, the proportion should not exceed 50 per cent (defined as a hen with one nude area of <5 cm in diameter)

There are several causes for higher mortality in the organic production.

Predation

Predation is the primary cause of mortality in around 12 per cent of the rearing facilities with organic pullets. Predation can both be ground predators, *i.e.* fox, and predator birds. The predation loss may be higher than actual reported, as when calculating the mortality based on the difference between the number of placed hens and slaughtered hens.

Cannibalism

Cannibalism is another common cause of mortality in systems with outdoor access, but is also a common cause in non-organic systems. Another management practice that affects mortality could be stocking density.

Helminth Infection

The organic production with access to free-range and a littered area enhances the risk of endoparasite infections, as they are potential infection sources. There are several endoparasites of poultry; however, not all are equally important for the commercial poultry production. *Ascaridia galli* and *Heterakis gallinarum* are the most common helminths. A helminth infection, despite of the level, will affect the welfare of the hens seen from the concept of biological function, as subclinical infections might increase morbidity and increase the severity of other diseases.

Some management practices affect the mortality. The choice of breed and genetic material may affect the mortality.

Even though the organic production of food has increased considerably over the last decade, it is still only of minor importance in poultry meat and egg production. There do not appear to be any insurmountable technical constraints as far as organic poultry production is concerned. There are, however, some areas of potential difficulty where attention to detail and good management is required, or where further development may be desirable:

- ☆ The suitability of breeds, in particular for table bird production, needs further consideration.
- ☆ The rearing of pullets organically from day-old chicks should be considered in preference to converting conventional pullets.
- ☆ Longer-term, the possibility of organic hatcheries should be investigated, particularly if less common breeds are advocated.
- ☆ Housing design, bedding and access to range should be appropriate to the bird's behavioral requirements - modifications to standards in line with Freedom Food recommendations could be considered.
- ☆ Flock size, stocking rates *etc.* need to correspond to a number of objectives, including stress minimization, integrating poultry in more diverse farming systems, and treating poultry as a land based activity - there needs to be a debate as to the best way to achieve this under organic production standards.

- More emphasis should be given to evaluating the contribution of vegetation, soil fauna and minerals as part of the contribution to poultry rations, as well as the potential of management practices to enhance soil fauna populations (*e.g.* earthworms in mulched vegetation).
- Major animal health problems are restricted to two issues: cannibalism and coccidiosis - cannibalism can be tackled using a combination of several management practices, while the emphasis on coccidiosis should shift from the permitted use of coccidiostats to management practices and the use of vaccines.
- Physical production performance is likely to be lower than for free-range poultry because of the wide range of management constraints imposed under current and proposed organic standards.

Marketing: Current Situation and Development Potential

There is the potential for the organic poultry sector to grow, both for meat and eggs, provided that:

- New producers secure a market before commencing production and that improved market opportunities increase confidence to enter/expand.
- Increased scale and consistency of supply and quality can encourage multiples to become involved, particularly given the general decline in the number of independent outlets, and the limited role of specialist organic retail, catering and mail-order outlets.
- Shared processing and packing facilities can be developed in appropriate locations to minimize transport distances and costs.
- Processing markets for outgrades (over/under-sized eggs, cracked eggs, layers, damaged table birds) can be developed.
- Pricing strategies do not discourage new customers from trying organic products.
- Improved consumer information and generic marketing of organic products helps to differentiate organic and alternative products such as free-range and barn-reared.
- Costs of labeling and certification are appropriate.
- Changes in production standards/legislation do not excessively add to production costs or technical feasibility of poultry production.

Chapter 8

Organic Aquaculture Production

Organic Management Plan

During registration of the farm by accredited Certification Body, the producer has to present an organic management plan to accredited Certification Body which requires to be verified during the inspection. This plan should be updated annually and shall apply to all aquatic organisms cultured in fresh and brackish water ponds and open water bodies in estuaries and sea (Black tiger shrimps, Indian major carps, freshwater prawns and bivalves) for production, processing and certification under these standards.

Adoption of organic aquaculture requires an interim period, 'the conversion period'. Commencement of the conversion period shall be taken as date of first inspection by the accredited Certification Body. The length of the conversion period would vary depending on the species, method of production, location and local conditions. Generally, for drainable systems where cleaning and disinfection is carried out, the conversion period shall be 6 months/one crop whichever is longer and in case of drainable and fallowed, the conversion period shall be 12 months. In case of non-drainable systems which can not be disinfected, the conversion period shall be 24 months (freshwater prawn, carps). In case of open water farming, the conversion period shall be considered as 3 months (bivalves).

Any kind of genetically engineered stock is prohibited. Stocks obtained through selective breeding are permitted, but seed production in this case should be based on organic principles. Collection of wild seed for selective stocking is prohibited (except for bivalves). Organically certified seed should be used. When organic seed is not available, the certifying body would prescribe a time limit for use of non-organic seed depending upon the species. For carps and fresh water prawns, the maximum percentage of non-organically produced juveniles allowed to be introduced to the farm shall be 80 per cent, 50 per cent and 0 per cent by second, third and fifth year from the year of notification.

The certified organic hatchery should source the initial stock from natural water bodies to raise them as brood stock with organic protocol at least for three months before their breeding. Synthetic hormone application for artificial propagation is not allowed. Since exogenous hormone supply is an essential requisite for induced spawning of carps, use of pituitary gland may be accepted. To avoid stress to the animal, thermal manipulation for accelerated larval development/growth or maturation, beyond natural range is prohibited in hatcheries. In carps, pre- and postponement of brood stock maturation through thermal/hormonal manipulation and their subsequent breeding is not permitted for seed production in certified organic hatchery.

For elimination of unwanted fishes, sun drying, netting or application of plant derivatives like tea seed cake (*Camellia sinensis*), mahua oil cake (*Bassia latifolia*), derris root powder (*Linchocarpus* sp.) and Neervalam (*Crotelaria tigilum*) are permitted in pond preparation. Use of any synthetic herbicides and pesticides are prohibited. Use of agricultural lime, dolomite or quick lime is permitted for disinfection and acidity corrections.

Stocking density to be limited so as not to compromise with the animal well being, ecological capacity of the site and species-specific physiological need and animal behavior. For shrimp farming, the maximum stocking density is 6 no.s/m2 and biomass in the pond shall not exceed 1400 kg/ha/crop and for freshwater prawns the stocking density up to 2.5 no.s/m2 and biomass in the pond shall not exceed 800 kg/ha/crop. For nursery rearing of freshwater prawns, the maximum stocking density of 20 no.s/sq.m is permitted. For carp fry and fingerlings production in nursery, the maximum stocking density is 2 million spawn/ha (200 no.s/m^2) and 0.1 million fry/ha (10 nos/m^2), respectively. For grow-out production of carps, maximum stocking density of 4,000 fingerlings/ha (0.4 nos/m^2) may be followed and the maximum biomass should not exceed 3 tonnes/ha at any point of time.

Pond Management

Ponds are required to be designed to maintain suitable environment most befitting with the natural behavior of the stock. Effluent water quality (nutrient load, suspended solids, ammonia *etc.*) has to be closely monitored at least twice in a crop (mid way and during harvest). In case of carp farming, floating vegetation cover with 10-15 per cent of the water surface should be provided in the production pond. Use of substrate for periphytic growth is permitted for enhancing the natural food availability in the pond. Use of plastics or any other synthetic materials may not be permitted for this purpose.

Supplementary Feeding

The natural feeding behavior of the animal should be explored to meet the nutritional and dietary need of the species for all its life stages. To meet requirements beyond the portion met by the natural productivity, certified organic feed should be provided. The non-organic feed is permitted only if organic feed is not available till initial one year of farming. Ingredients from Genetically Modified Organism (GMO) shall not be used. In case of tiger shrimp and freshwater prawn, the fish meal content in the feed should not exceed 20 per cent and the total protein content of animal origin should not exceed 25 per cent. In case of carp farming use of animal protein including fish meal in supplementary feed should be avoided. Feed prepared from certified organic ingredients avoiding possible entry of antibiotics/pesticides/ heavy metals/antioxidants/preservatives/growth hormones during the process is to be used for supplementary feeding. Excess feeding should be avoided. Growth promoters and synthetic amino acids are not permitted.

Health Management

Chemotherapeutics with allopathic veterinary drugs, and other harmful chemicals are prohibited. Herbal formulation and homeopathic medicines are allowed. Yeast based organic preparations and probiotics of certified origin are permitted to improve water/animal-rearing condition and to control pathogens. GMO based preparations are not permitted.

Approved List of Aquaculture Inputs

Piscicides of Herbal Origin	*Water/Soil Reformers/ Conditioners*	*Biofertilisers/ Manures/Nutrients (from organic sources)*	*Live Feed from Hatchery*
☆ Mahua Oil Cake (*Bassia latifolia*)	☆ Agri lime ($CaCO_3$)	☆ Compost from FYM	☆ Micro Algae
☆ Tea Seed Cake (*Camellia sinensis*)	☆ Quick Lime	☆ Vermi-Compost	☆ Artemia
☆ Neervalam (*Crotelaria tigilum*)		☆ Cowdung	☆ Moina
☆ Derris root powder (*Linchocarpus* sp.)		☆ Biodegradable processing by-products of animal/plant origin	☆ Branchionus
		☆ Micronutrients and essential chemical fertilizer for micro algal culture	☆ Copepodes
		☆ Mushroom spent wash	

List of Prohibited Antibiotics and Pharmalogically Active Substances for Aquaculture

1) Chloramphenicol
2) Nitrofurans including Furazolidone, Nitrofurazone, Furaltadone, Nitrofurantoin, Furylfuramide, Nifuratel, Nifursoxime, Nifurprazine and all their derivatives
3) Nemoycin
4) Nalidixic Acid
5) Sulphamethoxazole
6) *Aristolochia* spp. and preparations thereof

7) Chloroform
8) Cholrpromazine
9) Colchicine
10) Dapsone
11) Dimetridazole
12) Metronidazole
13) Ronidazole
14) Ipronidazole
15) Other nitroimidazoles
16) Clenbuterol
17) Diethylstilbestrol (DES)
18) Sulfonamide except approved sulfadimethoxine, sulfabromomethazine and sulfaethoxyrpyidazine
19) Floroquinolones
20) Glycopeptides

Organic Beekeeping/Apiculture

Choice of Breed/Strains

For the choice of bees for rearing, preference shall be given to indigenous species of bee, such as *Apis cerena indica, Apis mellifera, A.lorae, A.dorsata, Mellipona* spp. and *Trigona* spp.-Dammar (Indian stingless honey bees) and their local ecosystem.

Sources/Origin

The hives shall be made of the natural material to avoid contamination to the environment and the apiculture products. The bee wax for the new foundations shall be sourced from organic production units. Only natural products such as propolis, wax and plant oils shall be used in the hives.

Conversion Period

The conversion period shall not apply when bees are grown in wild and in natural conditions. One-year conversion period shall apply to those bee colonies/ apiaries which are reared. During conversion the bee colonies shall be placed in isolation and the foundation comb shall be made from organic wax.

Hiving the Honey Bees

The foundation comb shall be made from organic wax. For renovation of apiaries, 10 per cent per year of the queen bees and the swarms may be replaced by the non-organic queen bees and swarms in the organic production unit, provided that the queen bees and swarms are placed in hives with combs or comb foundations coming from organic production unit. Each bee hive shall primarily consist of natural materials. Use of construction materials with potentially toxic effects is prohibited. The apiaries shall be placed within a radius of 3 kms from the organic farms. These conditions shall not apply when the farms are not in flowering stage or when the hives are in the dormant condition.

Feed

The feeding of colonies shall be seen as an exception to overcome temporary feed shortages due to climatic conditions. It will only be permitted where the survival of the hives is endangered due to climatic conditions and only between the last honey harvest and 15 days before the start of the next nectar or honey dew flow period. The feed supplied shall be fully organic. Feeding shall be with organic honey, organic sugar syrup or organic sugar.

Management

For the renovation of apiaries, 10 per cent per year of the queen bees and swarms may be replaced by non-organic queen bees and swarms in the organic production unit provided that the queen bees and swarms are placed in hived with combs or comb foundations coming from organic sources. Clipping of wings of queen bees are prohibited.

Beehives shall be cleaned periodically. Each colony shall be periodically inspected once or twice in a month and in a manner causing least disturbance and provocation to honey bees.

Honey shall be extracted only from sealed combs. The used of brood combs is prohibited for honey extraction. At time of harvest, repellent consisting of prohibited substances (chemical synthetic repellents) shall not be used, except smoke. Excessive smoke shall not be used as it may taint the flavour of honey or otherwise spoil it. All the equipments used for extraction shall be thoroughly cleaned in boiling water, before use. During extraction, the honey shall run through a strainer of 1.40mm. The containers used for collecting the extracted honey shall be of stainless steel, aluminum or if of other metal, shall be thickly tinned or galvanized. The container shall have covers and each shall carry a label specifying the name of the producer, date and place of extraction. Honey extracted from the colonies with infectious bee diseases shall be kept separate and not mixed with general lot. This honey shall be pasteurized before marketing. It shall never be fed either in processed or unprocessed form to the bees. The extracted honey in air-tight containers shall be taken to the pooling and processing centres as early as possible.

Chapter 9

Advantages of Organic Farming

Organic agricultural practices are based on a maximum harmonious relationship with nature aiming at the non-destruction of the environment. The developed nations of the world are concerned about the spreading contamination of poisonous chemicals in food, feed, fodder and fiber. Naturally, organic farming system is looked upon as one of the means to remedy these maladies there.

Here are some advantages of organic farming described below:

1) Earth-Friendly Nature of Organic Farming

The relevance and need for an eco-friendly alternative farming system arose from the ill effects of the chemical farming practices adopted worldwide during the second half of the last century. Organic farming is considered to be the best among all of them because of its scientific approach and wider acceptance all over the world. In the Organic Trade Association's list of the ten best reasons to go organic, more than half have to do with the environment—with cleaner water supplies, more fertile soil, more sustainable farm practices, greater respect for a balanced ecosystem, and promote biodiversity among plants and animals. Organic growing methods also tend to bc friendlier to wildlife—birds, amphibians, earthworms, bees and other beneficial insects—in part by eliminating reliance on chemical pesticides, herbicides, fungicides, and fertilizers. These often not only eliminate organisms but degrade or eliminate their habitat. The antioxidant compounds are produced by plants in response to stress, such as that produced by insect pests or competing weeds. The need for these natural safeguards decreases with the use of herbicides and pesticides in conventional agriculture.

2) Mitigation of Global Warming through Global Warming

Organic production methods could mitigate climate change by reducing agricultural greenhouse gas (GHG) emissions, and increasing carbon (C) sequestration or curtailing further C loss into the atmosphere. The Fourth Assessment Report of the Intergovernmental Panel on Climate Change (IPCC) estimates that 10-12 per cent of global annual GHG emissions come from agriculture. Nitrous oxide (N_2O) accounts for 38 per cent of global GHG emissions from agriculture. Excessive N_2O emissions from the soil are due to high concentrations of mobile N from fertilizers and inefficient fertilizer application. The efficiency of synthetic fertilizers in conventional farming systems has steadily declined over time and increasing amounts of fertilizers are needed to maintain previous levels of productivity. Synthetic nitrogen fertilizer efficiency for cereal production has been declined by 50 per cent between 1960 and 2000. In this situation, organic farming may have a significant impact on slowing global warming. Soils rich in organic matter absorb and retain about fifteen to twenty-eight per cent more carbon, which is the equivalent of taking about 3,500 pounds of carbon dioxide per hectare out of the air. Organic farming could prove to be a significant and viable strategy for mitigating and adapting to climate change, especially for developing countries.

3) Improvement of Soil Quality in Organic Farming

Soil quality is the foundation on which organic farming is based. Soil building practices such as crop rotations, inter-cropping, symbiotic associations, cover crops, organic fertilizers and minimum tillage are central to organic practices. These encourage soil fauna and flora, improving soil formation and structure and creating more stable systems. In turn, nutrient and energy cycling is increased and the retentive abilities of the soil for nutrients and water are enhanced, compensating for the non-use of mineral fertilizers. Such management techniques play an important role in soil erosion control.

Efforts are directed to build and maintain the soil fertility through the farming practices. Multicropping, crop rotations, organic manures and pesticides, and minimum tillage are the methods employed for the purpose. Natural plant nutrients from green manures, farmyard manures, composts and plant residues build organic content in the soil. It is reported that soil under organic farming conditions has lower bulk density, higher water holding capacity, higher microbial biomass carbon and nitrogen and higher soil respiration activities as compared to the conventional farms. This indicates that sufficiently higher amounts of nutrients are made available to the crops due to enhanced microbial activity under organic farming.

Organically cultivated soils are relatively better attuned to withstand water stress and nutrient loss. Their potential to counter soil degradation is high and several experiments in arid areas reveal that organic farming may help to combat desertification. It is reported that about 70 hectares of desert in Egypt could be converted into fertile soil supporting livestock through organic and biodynamic practices. India, which has some areas of semi-arid and arid nature, can benefit from this idea.

Although it is generally accepted that all nutrients removed from the soil in crops should be replaced, modern non-organic agriculture usually only replaces the elements required for plant growth – nitrogen, phosphorus, potassium (these three being commonly referred to as 'NPK') and hence there is a continuing tendency in non-organic systems for trace nutrient deficiencies to develop and increase. This has been seen that the non-organically fertilized soils are generally equal or higher in P and K, but lower than compost enriched soils in organic matter in Ca, Mg, Mn, Cu and Zn.

Organic farming reduce N_2O emissions from agriculture by enhancing soil structure and microbial activity to increase N efficiency, and by encouraging the appropriate timing of nitrogen application to reduce leaching. Organically managed soils are also more aerated and have lower concentrations of mobile N, which further reduces N_2O emissions. Organic farming practices encourage the maximum recycling of on-farm nutrients. As a result, application rates of bio-fertilizers, when they are used, are often 60-70 per cent less than in conventional systems.

Besides, artificial fertilizers are not permitted as farmers are required to develop fertile soil by crop rotation and the use of compost, manure and clover. This delivers increased benefits to the environment when compared to the more intensive conventional methods of production.

4) Improvement of Soil Organic Matter (SOM)

Soil degradation caused by the use of synthetic fertilizers greatly impairs the accumulation of soil organic carbon (SOC). The potential for C sequestration is highly dependent on soil type and climate. Chemical fertilization and unsustainable land-use practices have degraded the world's agricultural soils. Conventional farming practices have caused extensive salinization and acidification. Topsoil loss due to soil erosion is one of the greatest threats to the future of agriculture. Organic farming builds up soil organic matter (SOM), which helps to avoid acidification, increases soil aggregate stability, and prevents erosion. Greater soil stability also increases water retention capacity, which reduces water logging and runoff. This lessens the need for irrigation. As a result, organic farms are shown to fare better than conventional farms during extreme weather events such as floods and droughts. Stable soils with high amounts of SOM also increase microbial and earthworm populations. Microbial activity is essential for nutrient cycling, enhancing plant-nutrient availability and acquisition. Microbial activity also increases soil aggregate stability, aeration and water percolation and helps protect plants against pathogens.

5) Better Water Quality in Organic Farming

As agriculture is dependent upon water availability, agricultural practices in turn impact water quality. Organic farming contributes to cleaner water by using biological fertilizers that release nutrients slowly, build soil organic matter, increase soil water-holding capacity and reduce leaching of nitrates into groundwater. The synthetic herbicides, pesticides, fungicides and fertilizers used in non- organic farming also seep into drinking water supplies, posing a variety of threats to human health. Only about 50 per cent of applied conventional fertilizers are taken up by crops; approximately 30-40 per cent is leached into ground and surface waters and the rest is lost into the atmosphere. Therefore, ground and surface waters are too often contaminated by pesticides, fertilizers and animal wastes that are not absorbed by plants or soil. Organic farming does not contaminate water to the degree that non-organic practices do because certified organic farmers do not use synthetic pesticides or highly soluble synthetic fertilizers.

6) Improvement of Water Holding Capacity of Soil in Organic Farming

Water availability is a particular concern around the globe as rainfall patterns have become increasingly unpredictable and groundwater use accelerates. Scarcity of clean, usable water is likely to become the largest problem facing global agricultural production in the near future. Organic agriculture has an advantage in this regard. It has been seen that increased soil organic matter under organic management also consistently measured greater water holding capacity in organically managed soils.

7) Organically Raised Animals are not given Antibiotics, Growth Hormones, or Fed Animal Byproducts

Feeding livestock animal byproducts increases the risk of mad cow disease (BSE) and several other diseases and the use of antibiotics can create antibiotic-resistant strains of bacteria. Organically-raised animals are given more space to move around and access to the outdoors, which help to keep them healthy.

8) Minimum Usage of Pesticides in Organic Farming

Organic cropping systems are designed differently and organic farmers in general tolerate higher levels of pests before they apply measures. In addition, most organic crops are more robust and rely less on the use of pesticides than conventional ones. Moreover, most pesticides are prohibited for use in organic production, and only some are allowed with restrictions. Due to the different management system it can be said in general that pesticide-related negative impacts are lower in organic than in conventional farms. Bio-control methods under organic farming are available in the country. In this regard, indigenous technological products such as Panchagavya (five products of cow origin), various leaf extracts have been found to control pest effectively.

9) Lower Growing Cost in Organic Farming

The economics of organic farming is characterized by increasing profits via reduced water use, lower expenditure on fertilizer and energy, and increased retention of topsoil. To add to this the increased demand for organic produce makes organic farming a profitable option for farmers.

10) Organic Farming Helps in Protection of Biodiversity

Organic farmers are both custodians and users of biodiversity at all levels. At the gene level, traditional and adapted seeds and breeds are preferred for their greater resistance to diseases and their resilience to climatic stress. At the species level, diverse combinations of plants and animals optimize nutrient and energy cycling for agricultural production. At the ecosystem level, the maintenance of natural areas within and around organic fields and absence of chemical inputs create suitable habitats for wildlife.

Biodiversity loss is a major threat to mankind. As around 1.5 billion hectares of the globe's land surface are used for crop production (FAO, 2003), agro ecosystems play an essential role in meeting this challenge. Conventional agriculture has largely contributed to the decrease in plant and animal biodiversity in agro-ecosystems through the loss of habitats and the heavy use of chemical inputs, namely pesticides and fertilizers. Alternative agricultural systems such as organic farming have been proven to be more favorable to biodiversity than conventional ones.

Table 9.1: Overview of the Relative Impacts of Organic Farming Compared with Conventional Farming

Area	*Aspect*	*Relative Environment. Impact**
Biodiversity	Floral diversity	+++
	Faunal diversity	++
	Habitat diversity	+
Landscape	Landscape structure and aesthetic value	+
Soil	Soil organic matter and acidity	++
	Soil structure	+?
	Soil biological activity	++
Ground and Surface Water	Nitrate leaching	++/-
	Phosphorus	+?
	Pesticides	+++
Climate and Air	Carbon dioxide (CO_2)	+?
	Nitrous oxide (N_2O)	+/-?
	Methane (CH_4)	+?
Energy	Intensity of energy use	++/-
	Efficiency of energy use	+?

(Adapted from Kasperczyk and Knickel, 2006).

* +: Slightly better; ++: Better; +++: Substantially better; ++/-: Better with some aspects that are negative; +?: Better with some uncertainties; +/-?: Partly better and partly worse with some uncertainties.

Organic fertilizing inputs and the use of compost supply the soil with organic matter which increases soil organic content, thus favoring detritivore organisms and the permanency of foodwebs. Beside plant nutrition, such processes are highly favorable to biodiversity and functional biodiversity. Tillage practices also contribute to soil aeration and are favorable to belowground living organisms which are highly favorable to an increase in biodiversity in agro-ecosystems at field, farm and landscape level. Therefore, positive effects of organic farming on species richness and diversity are especially visible in intensively-managed agricultural landscapes.

11) Efficient Utilization of Resource in Organic Farming

Future challenges for agriculture include changing energy systems, the globalization of food systems and a changing climate. With the impact of modern agricultural practices upon natural resources becoming increasingly critical, the question for policy-makers seeking to ensure a sustainable future production of food is how these challenges can best be addressed. Sustainable use of resources in agriculture is imperative if we are to overcome these challenges. A critical question in this regard is which types of agricultural systems, as well as practices, are most resource-efficient, in particular concerning energy and nutrient supply and use. Future energy challenges are based upon impending peak oil coupled with issues relating to climate change, whilst concerns regarding nutrient supply and use are based upon prognoses about limited phosphorus (and limited nitrogen availability determined by expensive energy sources) coupled with environmental concerns resulting from nutrient oversupply.

Table 9.2: The Overall Comparison between Organic and Inorganic Practice

Organic vs. Non-Organic	
Organic Produce	*Conventionally-Grown Produce*
Grown with natural fertilizers (manure, compost).	Grown with synthetic or chemical fertilizers.
Weeds are controlled naturally (crop rotation, hand weeding, mulching, and tilling).	Weeds are controlled with chemical herbicides.
Pests are controlled using natural methods (birds, insects, traps) and naturally-derived pesticides.	Pests are controlled with synthetic pesticides
Organic Meat, Dairy, Eggs	*Conventionally-Raised Meat, Dairy, Eggs*
Livestock are given all organic, hormone- and GMO-free feed.	Livestock are given growth hormones for faster growth, as well as non-organic, GMO feed.
Disease is prevented with natural methods such as clean housing, rotational grazing, and healthy diet.	Antibiotics and medications are used to prevent livestock disease.
Livestock must have access to the outdoors.	Livestock may or may not have access to the outdoors.

Future farming systems should thus seek to be highly resource-efficient, and profitable, whilst ensuring that practices are environmentally-sound and socially acceptable. Organic farming seeks to achieve economic, environmental and social dimensions of sustainability, and a fundamental objective of organic agriculture is to use renewable resources, increa se system recycling and reduce waste.

12) Restrictive Use of Food Additives and Ingredients in Organic Farming

Because of the restrictions on additive use required by organic food processing standards, a wide range and large quantity of potentially allergenic or harmful additives are avoided on a diet high in organically grown foods. Artificial food additives are widely used in the modern food processing industry. Many of them are relied on to put back the taste that processing often removes, to prevent spoilage or to improve the texture, color or flavor of foods. They are present in extraordinary abundance in the food supply, and several have been found to be unsafe after long use. Many are untested, and their widespread use continues despite questions about their safety. More than 500 food additives are permitted for use in non-organic processed foods. Food colorings such as tartrazine and other additives have been linked with symptoms such as allergic reactions, urticaria, headaches, asthma, growth retardation and hyperactivity in children.

In contrast, organic standards permit only a limited list of around 30 additives which are allowed in organic food products only if the authenticity of the product is respected and the product cannot be produced or preserved without them.

13) Employment Opportunities

Organic farming requires more labor input than the conventional farming system. Thus, India which has a very large amount of labor unemployment, under employment will find organic farming an attraction. Moreover, the problem of periodical unemployment will also get mitigated because of the diversification of the crops with their different planting and harvesting schedules resulting in the requirement of a relatively high labor input.

14) Rural Development

Gender equality is essential for sustainable rural development. Green Revolution technology and agricultural commercialization often serve to increase the share of female agricultural laborers, while constraining their opportunities for gainful employment. Agricultural mechanization has had the effect of pushing women out of economically productive tasks. At the same time, rising input costs increase the need for women to help on the farm as unpaid laborers. Moreover, women often lack access to land rights. Thus, although women play an essential role in agricultural production, they are often excluded from decision-making. Organic farming that adheres to the principle of fairness would promote gender equality. A diversified organic farming system could balance power relations by providing women with more meaningful tasks, therefore increasing income-earning opportunities and giving them a voice in production-related decisions.

15) Sustainability Over the Long Term

Organic agriculture considers the medium and long term effect of agricultural interventions on the agro-ecosystem. It aims to produce food while establishing an ecological balance to prevent soil fertility or pest problems.

Chapter 10

Health Effects of Organic Food

Food which is grown or raised through organic farming can have a major impact on physiological, mental and emotional health as well as the environment. Consequently, the choice of consuming organic foods is a growing trend in the world of today following the realization of healthy self improvement needs. However, finding exact facts about organic foods has always been confusing as people have varied opinions regarding its advantages as a whole. While some insights can be gained by investigating the nutritional components of food, the decisive test of the nutritional value of food is its capability to maintain health, growth and reproduction over successive generations of animals or humans. There are several factors which determine the nutritional value of organic food, including:

Physical Appearance

Shape, color, volume and firmness are some of the important visual characters of quality. Lot of comparisons has been made to ascertain the fact. Most of the studies indicate that practically there are no detectable differences between organically and conventionally grown fruits and vegetables. While shape and color are variety specific genetic characters, volume and firmness are dependent upon climate, environmental temperature and agricultural practices. However, under similar circumstances, products of conventional and organic systems are indistinguishable.

Taste

Taste, while not a direct influence on nutritional quality, can indirectly affect the nutritional value of a food through the stimulation of appetite and digestive processes. It is also an important quality distinction for consumers, and anecdotally many organic consumers feel that food grown organically tastes better. Organic consumers give this as a major reason for purchasing it. However, assessment of quality from taste point of view is difficult, as different persons have different standards depending upon their habits, liking and perception.

In general there is no evidence of better taste in organic fruits and vegetables. In vegetables where high water content is preferred organic vegetable may rate lower. But this is again subjected to individual preference and may not hold well under all circumstances. But certainly the flavoring ingredients, oils and other taste giving ingredients have been found to be higher in organic products.

It has been asserted that animal feed preference tests can reflect the quality of foods. Animals distinguish between the foods on offer from the various agricultural systems and almost exclusively prefer organic produce. The reason for this preference is not known. It may be due to the differences in the quality of food grown with organic as compared with mineral fertilizer. There is no doubt that the content of these quality differences should be fully established and the implications of such differences on human and animal nutrition should be assessed.

Although differences in taste are difficult to confirm in controlled studies, results of some studies are tabulated below:

Results	*Reference*
Better taste in organically grown peeled potatoes after four months storage	Dlouhy (1981)
Inconsistent and non-significant differences in the taste of different fruit and vegetables	Basker (1992)
Better total flavour strength, sweet taste and sugar content in organically grown carrots	Hogstad *et al.* (1993)
Taste and smell better in organically grown raw carrots and cabbage	Rembialkowska (2000)
Better taste in organic apples	Weibel *et al.* (2000)
Organically grown apples were sweeter and less tart	Reganold *et al.* (2001)

There are some possible reasons for a difference in taste between organically and non-organically grown produce which include:

- ☆ **Water content**: The higher water content in non-organically grown produce could dilute its flavor.
- ☆ **Variety**: Differences in choices of cultivars between the two systems might explain differences in taste.
- ☆ **Sugar content**: Organic produce has sometimes been shown to have a higher natural sugar content, which may be perceived as a better or worse taste, depending on the product.
- ☆ **Secondary nutrient content:** Levels of some aromatic compounds may be higher in organically grown produce.

However, there are number of difficulties with research on taste of organic food due to following listed factors including:

- ☆ **Cooking:** Taste differences are found in raw food but not in cooked food.
- ☆ **Ripeness**: Taste is more dependent on the level of ripeness or maturity of the samples (which is not fully controlled).

- ✰ **Subjectivity**: Taste is a personal thing and individual preferences can vary.
- ✰ **Preconceptions**: Consumers may have preconceived opinions about the superior taste of organic produce.

Nutritional difference between Organic and Conventionally grown Food

Organic standards are designed to protect and encourage soil micro-organisms, which are necessary to 'unlock' nutrients from the soil, form symbiotic relationships with plants and make nutrients available to them. Artificial fertilizers, herbicides and pesticides in conventional farming suppress these microorganisms, thus diminishing the 'life' of the soil and hindering the many complex and often poorly understood biological processes needed for nutrient supply to the plants. The possible reason for higher nutrient contents in organically grown crops may be due to soil biological activity. Therefore, the conversion period of two years required before a farm is awarded full organic status is partly to allow this soil microbiological life to recover and multiply. However, systematic reviews generally agree that the concentration of macronutrients, vitamins, and minerals in crops is either not at all or only slightly affected by the production system.

The following table describes the comparison of protein, nitrate and selected vitamin and mineral contents of organic *vs* conventionally grown crops (Worthington, 1998)

Nutrient	*Increased*	*Same*	*Decreased*
Protein quality	3	0	0
Nitrate	5	10	25
Vitamin C	21	12	3
β carotene	5	5	3
B vitamins	2	12	2
Ca	21	20	6
Mg	17	24	4
Fe	15	14	6
Zn	4	9	3

(No. of studies of organic crops shown to have increased, decreased or same nutrient content compared with conventionally grown crops)

Different studies may find very different results when measuring the same nutrient in conventional and organic crops. The reason for this variation between studies lies in the differing study designs, climatic conditions, soils, production years, crops, crop varieties, ripeness at harvest *etc.*, all of which may influence the nutrient content of a plant.

i) Higher Phenolic Compounds in Organic Foods

Modestly higher content of phenolic compounds has been found in organic food. Organic crops when compared to items grown through conventional methods have higher antioxidant activity and between 18 and 69 per cent higher concentrations of a range of individual antioxidants; including approximately 19 per cent higher levels of phenolic acids, 69 per cent higher levels of flavanones, 28 per cent higher levels of stilbenes, 26 per cent higher levels of flavones, 50 per cent higher levels of flavonols, and 51 per cent higher levels of anthocyanins. Increased intake of polyphenolics and antioxidants has been linked to a reduced risk of certain chronic diseases such as cardiovascular and neurodegenerative diseases and certain cancers.

ii) Higher Salicylic Acid in Organic Foods

Organic foods often contain higher levels of salicylic acid as compared to conventional one. This critical nutrient found in organic foods can help to proactively lower the risks of certain diseases for some individuals. Salicylic acid can help to keep a person's arteries from hardening due to the impact of heart disease and atherosclerosis.

iii) Higher Levels of Toxic Metals in Conventional Crops

Besides, conventional crops have higher levels of the toxic metal like cadmium, and are four-times more likely to contain detectable pesticide residues; there are general recommendations to minimize the intake of pesticides and cadmium to avoid potential negative health impacts.

iv) Higher Phytonutrient Compounds in Organic Crops

Phytonutrient compounds are synthesized by plants and do not appear to play a direct role in their growth and development, but rather in a diverse range of other plant processes including defence against herbivore and pathogen attack, protection against environmental stresses such as UV damage, and as attractants for pollinators. Organically grown crops will tend to have higher concentrations of phytonutrients than non-organically grown crops. Many of these compounds are produced by plants to protect themselves against attack, disease and damage; hence, if a plant is subjected to higher levels of stress it will accumulate higher levels of secondary compounds. Thus, if exogenous chemicals such as insecticides and fungicides are reduced or avoided, a greater reliance on the plants' own natural protection systems will result in the accumulation of higher levels of plant defence related compounds (secondary nutrients).

v) Nutritional difference between Organic and Non-organic Milk, Meat and Egg

For milk and dairy products, it has been demonstrated that organic products have a higher content of omega-3 PUFAs, across countries and seasons, due to a higher content of grass and roughage in the feed of organic cows. In a direct comparison, organic dairy products therefore have a more beneficial fatty acid composition. The same is apparently true for meat as well, although fewer studies have investigated this and therefore the evidence base is weaker. Dairy products make only a minor contribution to the omega-3 intake in humans. On average, replacing conventional with organic dairy products while keeping the diet constant will increase the intake of omega-3 PUFA by approximately 4 per cent. Replacing conventional meat products with organic meat products may increase the omega-3 intake by an additional 6 per cent, while approximately 50 per cent higher content of total omega-3 fatty acids (as per cent of total fatty acids) has been recorded in organic cow's milk as compared to conventional milk. Also, the content of ruminant FAs (a group of natural trans FAs produced in the cow's rumen) is higher in organic milk.

Organic milk contains higher levels of total conjugated linoleic acid (CLA), higher iron and α-tocopherol concentrations, which are all considered to be nutritionally desirable. CLA is a heart-healthy fatty acid with the potential of bolstering cardiovascular protection, and it is found in higher quantities in the meat and milk products of animals that have been pastured in free range. For eggs, it is likewise well described that the FA composition of the feed and consequently the access to pasture such as in organic systems, is a strong determinant of the fatty acid composition of the egg.

The preference of organic food is associated with a higher content of ruminant fatty acids in milk, which in turn is associated with a lower odds ratio for parent-reported eczema until age 2 years. A reduced risk of eczema in children fed only organic dairy products during infancy and whose mothers consumed only organic dairy products during pregnancy can be considered to provide stronger evidence of higher content of beneficial fatty acids in organic milk.

Conventional meat has slightly, but significantly higher concentrations of the saturated fatty acids myristic- and palmitic acid, which are linked to an increased risk of cardiovascular disease. Organic meat products are roughly 8 per cent higher in monounsaturated fat when compared to non-organic items. However, the content of saturated fatty acids, mono-unsaturated fatty acids and omega-6 PUFA is similar in organic and conventional milk. It should also be noted that several factors influence the fatty acid composition in milk. Policies are aiming at increasing the omega-3 intake of the population are likely to be far more efficient if they are directed at increasing the intake of omega-3- rich plant oils (*e.g.* rapeseed/canola and linseed) and fatty fish, with the desirable side effect of simultaneously reducing the intake of saturated fatty acids.

Controversial Disadvantages of Organic Food Consumption

i) Risk of Food Poisoning or Bacterial Infection

Food poisoning is caused by the presence of pathogenic bacteria in food items. Prominent among such pathogenic bacteria are *Salmonella, Campylobacter, Taenia soleum, Citrobacter freundii* and *Escherichia coli* strain 0157. All these bacteria are prevalent in animal guts and in their excreta. As organic cultivation relies on higher use of manures, it is assumed that they pose higher risk of contamination. However, after thorough evaluation and studies no such evidences have been found. Interestingly manures are recommended for all systems and conventional agriculture also promotes increased use of manures. Majority of the studies conclude that there is no risk of any food poisoning or bacterial infection through organic products. They are as safe as any other products produced by any other system.

ii) Threat of Mycotoxins/Phytotoxins in Organic Food

Certain moulds of fungi can produce phytotoxins which are harmful to human and animals. Some people have raised the concerns that as organic products are produced and stored without the use of fungicides, there are likely chances of such fungal contamination and resultant mycotoxins. All these concerns are sheer imaginations and there is not even a single study to substantiate the claim or concern. There is no evidence to suggest that consumption of organic foods has caused any mycotoxin/phytotoxin toxicity and on this issue they are at par with non-organic products.

Study on Effect of Organic Food on Human Health

Current research on the role of organic food consumption in human health is scarce, as compared to other nutritional epidemiology topics. Long-term studies aiming to identify potential links between organic food consumption and health are lacking. In particular, long-term interventional studies aiming to identify potential links between organic food consumption and health are underprovided, mainly due to high costs. Prospective cohort studies constitute a feasible way of examining such relationships, although compliance assessment is challenging. Considering a lack of biomarkers of exposure, the evaluation of the exposure, *i.e.* organic food consumption, will necessarily be based on self-reported data that may be prone to measurement error. Taking into account the difficulties of performing long-term dietary intervention studies in humans, animal studies offer some potential of studying long-term health effects of foods *in vivo.* However, extrapolation of the results from animal studies to humans is not straightforward. A growing literature is aiming at characterizing individual lifestyles, motivations and dietary patterns in regard to organic food consumption, which is generally defined from responses obtained from food frequency questionnaires. Some recent researches from clinical studies have been compiled to find the association between consumption of organic food and health.

Health Benefits Linked to Organic Food Consumption

Organic food is extremely popular for its known health benefits. However, the relationship between the nutritional value of a product and health is difficult to predict. Although numerous studies have compared the nutrient, antioxidant, and pesticide residue content of organic and conventional foods, few scientific studies in animals or humans have examined whether the consumption of organic food is associated with better health than consuming the corresponding conventional food.

i) Atopy, Allergies, Eczema and/or Wheeze Occurrence

It has been found that organic food has protective role against atopy, allergies, eczema and/or wheeze occurrence. Organic products contain on average more antioxidants, especially vitamin C, and organic dairy products contain on average a higher amount of conjugated linoleic acid and n-3 fatty acids. A preventive effect of antioxidants on wheezing complaints has been suggested possibly through protection of the epithelium of the bronchial tubes against oxidative stress. Further, n-3 fatty acids might have anti-inflammatory properties and can contribute to the skin barrier, thus potentially decreasing atopic reactions. Organic dairy products also contain more gram-negative bacteria and higher levels of the corresponding lipopolysaccharide than conventional, pasteurized dairy products. Therefore, the protective factor associated with the consumption of organic dairy products could be the ingestion of non-infectious microbial components with a possible effect on gut immunity development.

ii) Nutritional Cancer Therapies

Three pesticides frequently used in agriculture *i.e.* glyphosate, malathion, and diazinon—act as carcinogenic to humans based on evidence from studies of occupational exposure in humans and laboratory studies in animals. Through occupational exposure (primarily in agricultural settings), malathion is associated with prostate cancer, diazinon is associated with lung cancer, and all 3 pesticides are individually linked to non-Hodgkin lymphoma. In the general population, low-level pesticide exposure is widespread, and the primary route of exposure is diet, especially through intake of conventionally grown fruits and vegetables. Organic foods are produced without synthetic pesticides and are less likely to contain pesticide residues than conventionally produced, nonorganic foods. It has been shown that switching from consuming conventionally grown foods to organic foods decreases urinary concentrations of pesticide metabolites, suggesting reduced exposure to pesticides. More recent clinical evidence comes from doctors and nutritionists administering 'alternative' cancer treatments who have observed that a completely organic diet is essential for a successful outcome. Nutritional cancer therapies involve avoidance of pollutants and toxins as much as possible, therefore, the exclusive consumption of organically grown foods increases in nutrient intakes, and have yielded good results.

iii) Cardiovascular Disease and Hypertension

It has been seen that, organic food consumers (occasional and regular), as compared to non-consumers, exhibited a lower incidence of cardiovascular disease. Organic foods are PUFA rich; therefore replacing SFA-rich foods by PUFA-rich foods may be the reason to decrease the risk of cardiovascular diseases. Besides, the ingestion of organic food results in a significant reduction of the key-marker levels related to cardiovascular diseases, such as: total cholesterol, LDL, glucose, monocytes, and lymphocytes. These markers are related to damages in the vascular endothelium and in the formation of the atherosclerotic plaque, a condition that can cause hypertension and cardiovascular diseases.

iv) Type-2 Diabetes

With data from several scientific researches, an inverse association has been established between the consumption of organic foods and diabetes. Regular consumption of organic foods is found to be associated with lower odds of type 2 diabetes. Organic animal foods or products are known to contain reduced amounts of antibiotic residues compared to conventional animal foods or products. Chronic antibiotic exposure is associated with a higher risk of type 2 diabetes via possible mechanisms including antibiotic-driven changes in insulin sensitivity, glucose tolerance, lipid deposition, and energy harvesting potential by altering the gut microbiota composition. Furthermore, organic animal products may contain less sex steroids or growth hormones used in farm animals compared to conventional animal products. Lastly, organic foods such as eggs, milk, and meats may contain more beneficial unsaturated fatty acids such as omega-3 fatty acids than conventional products. These fatty acids could have beneficial effects on insulin sensitivity and is likely to reduce the risk of diabetes.

v) Pre-eclampsia

It has been reported that women who are eating organically grown vegetables 'often' or 'mostly' have lower risk of developing pre-eclampsia than women who are not consuming organic vegetables or to do so less frequently. The possible explanations are based on the following characteristics of organically produced vegetables rather than conventional vegetables: (1) lower dietary pesticide exposure; (2) higher intake of secondary plant metabolites and (3) possibly a different microflora on organic vegetables, which could affect human (maternal) intestinal microbiota in a beneficial way. Furthermore, pesticides, or the absence of them, might impact the composition of the gut microbiota.

vi) Semen Quality

Studies have indicated that exposures to pesticides either dietarily through conventional food or occupationally may adversely influence sperm quality. However, very few studies have been conducted in human. Nonetheless, controlled studies with animals have given some strong relationship of superior sperm quality with the foods from organic farming.

vii) Hypospadias

Hypospadias and cryptorchidism are both related to androgen receptor function. Numerous pesticides are known to be endocrine-active substances with estrogenic or anti-androgenic activity. Dietary pesticide exposure through conventional food could hypothetically lead to hypospadias via hormonal or placental disturbances. Placental insufficiency may play an important role in hypospadias etiology as fetal growth restriction and preeclampsia both are risk factors for hypospadias. It has been seen in few scientific studies that placental function may link the lower prevalence of hypospadias with organic vegetable consumption.

viii) Stronger Immune System

The traditional or industrial farming practices aim at enhancing production and farm output by all necessary means, for example, the notion of producing more cereals, more meat and bigger fruits through genetic modifications. Use of growth hormones seems to solve some of the world's food insecurity concerns. The effects are not yet visible, but in the long-term, the consequences are sensitivity to allergens and a major reduction in immune system strength.

By eating organic foods, the risks of decline in immune system strength are significantly reduced because organic foods are not altered at all. Furthermore, organic foods have quality and higher vitamin and mineral contents that help to strengthen the human immune system.

Other Positive Health Effects

There are several other positive health effects reported in humans resulting from the consumption of organically grown foods. Some of those are listed below:

- ✰ Lower incidences of catarrhal conditions
- ✰ A "very marked decline" in colds and influenza
- ✰ More rapid convalescence
- ✰ Excellent health
- ✰ Fewer sports injuries in sportsman
- ✰ Greater resilience to fractures and sprains
- ✰ Clear and healthy skin
- ✰ Improved dental health
- ✰ Reduction in risk of obesity.

Organically Grown Food and Animal Health

There have been numerous studies published comparing the health of animals on organically or non-organically grown feed, and it is not possible to confirm the authenticity of the organically and non-organically grown feeds compared in many tests. Therefore some of the health effects established from different scientific research have been presented below.

Rabbit

Higher pregnancy rates, more embryos, larger litters and better health of rabbits

Chicken

Better weight gain after coccidial illness and fewer incidents of illness in chickens fed organically grown feed; significantly faster growth, higher egg weight, yolk weight and body weight at 32 weeks in organically fed chickens compared with birds fed chemically fertilized feed.

Livestock

Organically produced crops may be more conducive to good health for livestock also. The positive health effects are most striking in sick or otherwise vulnerable animals such as newborns and in sensitive areas of reproduction such as sperm motility. It is particularly interesting to see that the fertility of animals fed fodder grown with chemical fertilizers and pesticides declined over several generations.

It is tempting to conclude that (except for iodine intake with milk) organic food consumption results in higher dietary intakes of a range of nutritionally desirable compounds such as antioxidants, certain vitamins, and omega-3 fatty acids, but lower intakes of nutritionally undesirable pesticides, cadmium, and saturated fatty acids. However, an anthroposophic lifestyle differs with regard to many characteristics, including regular consumption of fermented vegetables, which may influence the gut microbiota, and the effects observed in studies involving families with this lifestyle may not be associated with organic food consumption alone.

Chapter 11

Process of Certification of Organic Farming

The need for a trustworthy and healthy organic market is important as the consumers are becoming increasingly separated from the food production process. The void is filled in by a plethora of processing, distribution and marketing intermediaries. The consumers should possess knowledge and confidence that the products they buy are truly organic. Such environment can be created by an organized system of inspection and certification supported by law and regulations. Organic products enjoy trust based on the label or logo approved by the certification process.

Organic certification addresses a growing worldwide demand for organic food. It is intended to assure quality and prevent fraud. For organic producers, certification identifies suppliers of products approved for use in certified operations. For consumers, "certified organic" serves as a product assurance, similar to "low fat", "100 per cent whole wheat", or "no artificial preservatives". Certification is essentially aimed at regulating and facilitating the sale of organic products to consumers. Individual certification bodies have their own service marks, which can act as branding to consumers. Most certification bodies operate organic standards that meet the national government's minimum requirements.

Organic Certification

It is a certification process for producers of organic food and other organic agricultural products. In general, any business directly involved in food production can be certified, including seed suppliers, farmers, food processors, retailers and restaurants. Requirements vary from country to country, and generally involve a set of production standards for growing, storage, processing, packaging and shipping that include:

- ☆ Avoidance of synthetic chemical inputs (*e.g.* fertilizer, pesticides, antibiotics, food additives, *etc.*) and genetically modified organisms.
- ☆ Use of farmland that has been free from chemicals for a number of years (often, three or more).
- ☆ Keeping detailed written production and sales records (audit trail).
- ☆ Maintaining strict physical separation of organic products from non-certified products.
- ☆ Undergoing periodic on-site inspections.

Purpose of Certification

A label on an organic product conveys that the manufacturer has a license for organic production, an independent agency has inspected the production/processing practices followed by the producer, and compliance of the set standards is made. Such an assurance becomes crucial in the generation of consumer confidence in the organic products, particularly as they are costlier than the conventional ones. The marketing agents of organic products also need such an assurance to maintain fair trade practices essential to the growth and sustainability of the demand for such products. Certification is necessary for farmers to access organic export markets and to ensure that they receive a higher price for their products on domestic markets. The aim of organic certification is, therefore, to (a) provide an identifiable label (a logo or a symbol) and (b) give an assurance to the consumers that the product is truly organic.

However, third-party certification can be a complicated and costly process for farmers. It would have been more appropriate to compliment and reinforce sustainable aspects of the existing farming system and to establish a local certification body based on local standards.

General Requirements for Certification

A person seeking to receive or maintain organic certification under the regulations in this part must establish, implement, and update annually an organic production or handling system plan that is submitted to an accredited certifying agent and permit on-site inspections with complete access to the production or handling operation, including noncertified production and handling areas, structures, and offices by the certifying agent. He or she should maintain all records applicable to the organic operation for not less than 5 years beyond their creation and allow authorized representatives of the Secretary, the applicable State organic program's governing State official, and the certifying agent access to such records during normal business hours for review. The person should submit the applicable fees charged by the certifying agent.

Application for Certification

A person seeking certification of a production or handling operation must submit an application for certification to a certifying agent. The application must include an organic production or handling system plan, the name of the person completing the application; the applicant's business name, address, and telephone number; and, when the applicant is a corporation, the name, address, and telephone number of the person authorized to act on the applicant's behalf.

Review of Application

Upon acceptance of an application for certification, a certifying agent must review the application to ensure completeness, determined by a review of the application materials whether the applicant appears to comply or may be able to comply with the applicable requirements, verify that an applicant who previously applied to another certifying agent and received a notification of noncompliance or denial of certification, has submitted documentation to support the correction of any non-compliances identified in the notification of noncompliance or denial of certification and schedule an on-site inspection of the operation to determine whether the applicant qualifies for certification if the review of application materials reveals that the production or handling operation may be in compliance with the applicable requirements.

The certifying agent shall within a reasonable time review the application materials received and communicate its findings to the applicant, provide the applicant with a copy of the on-site inspection report, as approved by the certifying agent, for any on-site inspection performed; and provide the applicant with a copy of the test results for any samples taken by an inspector.

The applicant may withdraw its application at any time. An applicant who withdraws its application shall be liable for the costs of services provided up to the time of withdrawal of its application. An applicant that voluntarily withdraws its application prior to the issuance of a notice of noncompliance will not be issued a notice of noncompliance. Similarly, an applicant that voluntarily withdraws its application prior to the issuance of a notice of certification denial will not be issued a notice of certification denial.

On-site Inspections

A certifying agent must conduct an initial on-site inspection of each production unit, facility, and site that produces or handles organic products and that is included in an operation for which certification is requested. An on-site inspection shall be conducted annually thereafter for each certified operation that produces or handles organic products for the purpose of determining whether to approve the request for certification or whether the certification of the operation should continue. The initial on-site inspection must be conducted within a reasonable time following a determination that the applicant appears to comply or may be able to comply with the requirements. All on-site inspections must be conducted when an authorized representative of the operation who is knowledgeable about the operation is present and at a time when land, facilities, and activities that demonstrate the operation's compliance with or capability to comply with the applicable provisions can be observed.

The on-site inspection of an operation must verify that prohibited substances have not been and are not being applied to the operation through means which, at the discretion of the certifying agent, may include the collection and testing of soil; water; waste; seeds; plant tissue; and plant, animal, and processed products samples.

The inspector must conduct an exit interview with an authorized representative of the operation who is knowledgeable about the inspected operation to confirm the accuracy and completeness of inspection observations and information gathered during the on-site inspection. The inspector must also address the need for any additional information as well as any issues of concern.

At the time of the inspection, the inspector shall provide the operation's authorized representative with a receipt for any samples taken by the inspector. There shall be no charge to the inspector for the samples taken. A copy of the on-site inspection report and any test results will be sent to the inspected operation by the certifying agent.

Granting Certification

Within a reasonable time after completion of the initial on-site inspection, a certifying agent must review the on-site inspection report, the results of any analyses for substances conducted, and any additional information requested from or supplied by the applicant. If the certifying agent determines that the organic system plan and all procedures and activities of the applicant's operation are in compliance with the requirements and that the applicant is able to conduct operations in accordance with the plan, the agent shall grant certification. The certification may include requirements for the correction of minor non-compliances within a specified time period as a condition of continued certification. The certifying agent must issue a certificate of organic operation which specifies the name and address of the certified operation; effective date of certification; categories of organic operation, including crops, wild crops, livestock, or processed products produced by the certified operation; and name, address, and telephone number of the certifying agent. Once certified, a production or handling operation's organic certification continues in effect until surrendered by the organic operation or suspended or revoked by the certifying agent, the state organic program's governing state official, or the administrator.

Denial of Certification

When the certifying agent has reason to believe, based on a review of the information, that an applicant for certification is not able to comply or is not in compliance with the requirements, the certifying agent must provide a written notification of noncompliance to the applicant. When correction of a noncompliance is not possible, a notification of noncompliance and a notification of denial of certification may be combined in one notification. The notification of noncompliance shall provide a description of each noncompliance; the facts upon which the notification of noncompliance is based; and the date by which the applicant must rebut or correct each noncompliance and submit supporting documentation of each such correction when correction is possible.

Upon receipt of such notification of noncompliance, the applicant may correct non-compliances and submit a description of the corrective actions taken with supporting documentation to the certifying agent; or correct non-compliances and submit a new application to another certifying agent, provided that, the applicant must include a complete application, the notification of noncompliance received from the first certifying agent, and a description of the corrective actions taken with supporting documentation.

Continuation of Certification

To continue certification, a certified operation must annually pay the certification fees and submit the following information, as applicable, to the certifying agent:

1) An updated organic production or handling system plan which includes:
   (i) A summary statement, supported by documentation, detailing any deviations from, changes to, modifications to, or other amendments made to the previous year's organic system plan during the previous year; and
   (ii) Any additions or deletions to the previous year's organic system plan, intended to be undertaken in the coming year
2) Any additions to or deletions from the information required
3) An update on the correction of minor no-compliances previously identified by the certifying agent as requiring correction for continued certification; and
4) Other information as deemed necessary by the certifying agent

Following the receipt of the information, the certifying agent shall within a reasonable time arrange and conduct an on-site inspection of the certified operation. The certifying agent may allow continuation of certification and issue an updated certificate of organic operation on the basis of the information submitted and the most recent on-site inspection conducted during the previous 12 months.

Fees and other Charges for Certification

Fees charged by a certifying agent must be reasonable, and a certifying agent shall charge applicants for certification and certified production and handling operations only those fees and charges that it has filed with the administrator. The certifying agent shall provide each applicant with an estimate of the total cost of certification and an estimate of the annual cost of updating the certification. The certifying agent may require applicants for certification to pay at the time of application a nonrefundable fee which shall be applied to the applicant's fees-for-service account. The certifying agent may set the nonrefundable portion of certification fees; however, the nonrefundable portion of certification fees must be explained in the fee schedule submitted to the administrator. The fee schedule must explain what fee amounts are nonrefundable and at what stage during the certification process fees become nonrefundable. The certifying agent shall provide all persons inquiring about the application process with a copy of its fee schedule.

What has to be Certified?

Each production or handling operation or specified portion of a production or handling operation that produces or handles crops, livestock, livestock products, or other agricultural products that are intended to be sold, labeled, or represented as "100 per cent organic," "organic," or "made with organic (specified ingredients or food group(s))" must be certified. Any operation that knowingly sells or labels a product as organic, except in accordance with the Act, shall be subject to a civil penalty.

Record Keeping by Certified Operations

A certified operation must maintain records concerning the production, harvesting, and handling of agricultural products that are or that are intended to be sold, labeled, or represented as "100 per cent organic," "organic," or "made with organic (specified ingredients or food group(s))." Such records must be adapted to the particular business that the certified operation is conducting. It should fully disclose all activities and transactions of the certified operation in sufficient detail as to be readily understood and audited. It should be maintained for not less than 5 years beyond their creation.

Accrediting Agencies for Certification

Organic production requires certification after periodic inspections in order to ensure that all prescribed practices are followed. The inspection and certification are done by the agencies accredited to the Accrediting Agencies dealing with the commodity. Inspection and certification agencies can be government departments, NGOs, trade or consumer or producer organizations. Such agencies should be registered bodies, with managements in position, declaring the persons who shall be held responsible for any miscarriage of certification and having proof of adequate field staff to undertake periodic inspections. The continued accreditation of such bodies is dependent upon their record of fidelity to the principles of organic production. They are authorized to award certificates after due satisfaction that practices conformed to those enunciated by the Accrediting Agency in relation to the item concerned. The charges levied by the certifying agencies are fixed by the Accrediting Agencies. Certification of products is the foundation on which organic agriculture rests. Its role in production and marketing of organic foods becomes a significant corner stone in the creation of a confident consumer community. The organic standards have the dual function of providing guaranties to consumers that production rules have been adhered to as well as giving guidance and advice to the producers on how organic principles are to be applied on the farm, with the certification bodies acting as intermediaries between these two interest groups. Standards also act as the basis for the contract between producer and certification body, which is binding on both parties.

International Federation of Organic Agriculture Movements (IFOAM)

IFOAM is the acronym for the International Federation of Organic Agriculture

Movements, the umbrella organization for organic movements worldwide. One of its most important tasks is to set the international standards for what may be labeled as "organic". IFOAM also coordinates the international network of organic organizations and represents, internationally, the organic movement in parliamentary, administrative and policy-making forums. The IFOAM standards are revised every other year (from 2003 every third year) and are ratified by the IFOAM General Assembly. They are implemented locally by national or regional certification and inspection organizations that adopt their own standards within the framework set by the Basic Standards. Some organic organizations used standards long before IFOAM adopted standards. The first legislation on organic farming was already adopted by the states of Oregon and California in the 1970's, but most countries waited much longer.

American national regulations came into effect October 2002 (National Organic Program; AMS-USDA, 2000). On the international level, EU regulations for plant production were introduced in 1991 (Council Regulation, 1991) and for livestock in 1999 (Council Regulation, 1999). The FAO/WHO Codex Alimentarius guidelines were adopted for organic plant production in 1999 and for livestock in 2001 (FAO, 2001).

The main functions of organic farming were laid down by IFOAM in 1992.

- To produce food of high nutritional quality in sufficient quantity
- To interact in a constructive and life enhancing way with all natural systems and cycles
- To encourage and enhance biological cycles within the farming system, involving micro-organisms, soil flora and fauna, plants and animals
- To maintain and increase long term fertility of soils
- To use, as far as possible, renewable resources in locally organized agricultural systems
- To work, as far as possible, within a closed system with regard to organic matter and nutrient elements. This aims to reduce external inputs
- To work, as far as possible, with materials and substances which can be reused or recycled, either on the farm or elsewhere
- To give all livestock living conditions which will allow them to perform the basic aspects of their innate behavior
- To minimize all forms of pollution that may result from agricultural practices
- To maintain the genetic diversity of the agricultural system and its surroundings, including the protection of plant and wildlife habitats
- To consider the wider social and ecological impact of the farming system

Participatory Guarantee Systems (PGS)

For smooth functioning of IFOAM, it is supporting the development of Participatory Guarantee Systems (PGS). As defined by IFOAM, "Participatory

Guarantee Systems are locally focused quality assurance systems. They certify producers based on active participation of stakeholders and are built on a foundation of trust, social networks and knowledge exchange." PGS is an approach based on the honor system, whereby standards and compliance are set and monitored by local consumers and organic producers. Products are intended for local and domestic markets. PGS standards are context-specific, varying with the local cultural practices, ecologies, markets and politics. PGS initiatives are serving thousands of small organic farmers and their consumers all over the world, and the numbers are increasing every year.

International Standards

The International Federation of Organic Agriculture Movements (IFOAM) has produced a set of international organic standards, laid down by people from many countries. These give guidelines about what organic farming is and how it should be practiced on the farm. International standards are also used to help countries set their own standards, which take into account different farming systems. Many countries have an organic standards authority which lays down national standards and awards a symbol to farms which have followed the standards. This symbol then allows farmers to market certified organic produce. This is important, as it ensures that people know that the food which they buy is organic.

Regulatory Mechanism of Organic Production in India

In India, most of the farms are 'organic by default' (excluding conventional). Today, farmers need to have a certification to refer their farm practice as organic and products are organic products in the market. The actual data, percentage of organic farmland in Indian continent are unknown and the majority of the farmers, tribal people are non-certified organic farmers, as they are engaged in indigenous farming. This farm practice has gained attention among farmers, entrepreneurs, policy makers and agricultural scientists as it reduces the external expenditure on farm materials, farmers' dependency on financial loans and decreases in cost of production. The high demands of organic consumers in international and national markets and development of fair trade markets have increased the agricultural profit of the farmers in the organic market. The development of Indian organic farming is associated with the expansion of agro-technologically driven sustainable farming promoted by agriculture practitioners, scientists and academicians. The acceptance of organic farming as an official socio-political discourse in the central government development perspective is a very recent one. The country is promoting a holistic discourse in organic farming by encouraging farmers to construct values and develop an interest in agriculture in general and organic farming in particular. Different agencies have prompted organic agriculture in the country; but government initiatives did not materialize in organic farming until its eighth five-year plan.

The Government of India took a serious interest in organic farming when European markets rejected the food consignments; due to the unacceptable level of pesticides found in the Indian export commodities. Government of India, considering the constraints of the export market, showed interest in recognizing organic farming and, certified organic farms for export purpose. Interestingly, the Ministry of Commerce also took an interest in the concept of the organic farming. On the other hand, the Ministry of Agriculture, which deals directly with different agricultural practice, farmers did not bother to explore the possibility of introducing organic farming practice as an experiment, till recently. The year 2000 is a landmark year in Indian organic farming development as central government took four major decisions as pointed by Bhattacharyya and Chakraborty (2005; 2008),

1) The Planning Commission constituted (2000) a National Steering Committee (NSC) that identified implementation of organic farming as a National challenge and suggested introduction of organic farming through projects as a major thrust area for the 10th Plan. It recommended the introduction of organic farming in the North East Region and rain fed areas, where the consumption of agro chemicals is low or negligible.
2) The National Agricultural Policy (2000) recommended promotion of traditional knowledge of agriculture relating to organic farming and it is scientific upgradation.
3) The Department of Agriculture and Cooperation (DAC), Ministry of Agriculture constituted (2000) a Taskforce on organic farming and recommended promotion of organic farming.
4) The Ministry of Commerce launched the National Organic Programme in April 2000.

The organic farming policy has introduced standards and certifications for the organic farming practice. The civil society or service providers of organic farming were given the opportunity to apply standards at the field level to reduce the technical difficulties during certification. The policy has integrated organic farm approaches within the conventional food system, as the policy did not make it mandatory to use traditional seeds in organic farming practice. In India, organic farming has higher demand, due to its diverse climatic conditions and provides an opportunity for the production of diverse food products as these have greater demand in the international market.

In India, there are two accreditation systems for authorizing Certification and Inspection agencies for organic certification. National Programme on organic Production (NPOP) promoted by Ministry of Commerce is the core programme which governs and defines the standards and implementing procedures. National Accreditation Body (NAB) is the apex decision making body. Certification and Inspection agencies accredited by NAB are authorized to undertake certification process. The NPOP notified under Foreign Trade Development and Regulation Act (FTDR) act and controlled by Agricultural Processed Foods Export Development Authority (APEDA) looks after the requirement of export while NPOP notified under Agriculture Produce Grading, Marking and Certification (APGMC) act and controlled by Agriculture Marketing Advisor (AMA), Directorate of Marketing and Inspection looks after domestic certification. In 2006, India's organic certification process under NPOP has been granted equivalence with European Union and Switzerland. It has also been recognized for conformity assessment by USDA's NOP.

National Programme for Organic Production (NPOP)

In 2001, the Government of India established a National Steering Committee for Organic Production (NSCOP) under the Department of Commerce, which launched the National Program on Organic Production (NPOP). The NPOP has established the National Standards for Organic Production (NSOP), which was recognized as equivalent to US and EU standards in 2006. The NPOP has also set up a certification accreditation program in accordance with these standards (FAO 2011). The NPOP offers extension services to provide technical and financial support for organic farmers during conversion; however these services are quite limited in size and scope (FAO 2011). India also has a well-established PGS certification program. For a farm to receive full organic certification, which is legally required before labeling any produce as organic, it must undergo a two year conversion period, then is deemed fully organic from the third year (by Soil Association standards). This is to allow the development of soil fertility and a reduction in residual pesticides.

National Program on Organic Production (NPOP) was launched during 2001 under the Foreign Trade and Development Act (FTDR Act). The document provides information on standards for organic production, systems criteria, and procedures for accreditation of Inspection and Certification bodies, the national organic logo and the regulations governing its use.

Scope

The NPOP shall, among others, include: (i) Policies for development and certification of organic products, (ii) National standards for organic products and processes, (iii) Accreditation of programmes to be operated by Inspection and Certification Agencies and (iv) Certification of organic products

Operational Structure

National Steering Committee for National Programme for Organic Production, is the apex policy making body and operates the entire programme through National Accreditation Body (NAB), Technical Committee (TC) and Evaluation Committee (EC).

Agricultural and Processed Food Products Export Development Authority (APEDA) is the secretariat and implementation office for NPOP for export while Agriculture Marketing Advisor, Directorate of Marketing and Inspection, Department of Agriculture and Cooperation is the secretariat and implementation office for NPOP for domestic certification.

National Standards for Organic Production (NSOP)

In India, standards for organic agriculture were announced in May, 2001 and the National Programme on Organic Production (NPOP) is administered by Agricultural and Processed food products Export Development Authority (APEDA) using the IFOAM basic standards under the Ministry of Commerce. It includes definite principles, basic standards of production, documentation, inspection and certification guidelines. The Government has set the frame conditions in which the organic sector of a country operator which include content and legal status of organic standards, the regulations concerning the use of organic claims and labels, the legislation on consumer protection and the accreditation system.

As per the national accreditation policy, all certifying agencies operating in India are to obtain accreditation from any one of the accrediting agency appointed by the Government of India (*E.g.* Spice Board, Coffee Board, Tea Board, APEDA, *etc.*). National Standards for Organic Production are grouped under following six categories:

1) Conversion
2) Crop production
3) Animal husbandry
4) Food processing and handling
5) Labeling
6) Storage and transport

Standard requirements for crop production, food processing and handling are listed below:

1) Conversion Requirements

The time between the start of organic management and cultivation of crops or animal husbandry is known as the conversion period. All standard requirements should be met during conversion period. Full conversion period is not required where organic farming practices are already in use.

2) Crop Production

Choice of Crops and Varieties

All seeds and planting materials should be certified organic. If certified organic seed or planting material is not available then chemically untreated conventional material can be used. Use of genetically engineered seeds, pollen, transgenic plants are not allowed.

Duration of Conversion Period

The minimum conversion period for plant products, produced annually is 12 months prior to the start of the production cycle. For perennial plants (excluding pastures and meadows) the conversion period is 18 months from the date of starting organic management. Depending upon the past use of the land and ecological situations, the certification agency can extend or reduce the minimum conversion period.

Fertilization Policy

Biodegradable material of plant or animal origin produced on organic farms should form the basis of the fertilization policy. Fertilization management should minimize nutrient losses, avoid accumulation of heavy metals and maintain the soil pH. Emphasis should be given to generate and use own onfarm organic fertilizers. Brought in fertilizers of biological origin should be supplementary and not a replacement. Over manuring should be avoided. Manures containing human excreta should not be used on vegetation for human consumption.

Pest Disease and Weed Management including Growth Regulators

Weeds, pests and diseases should be controlled preferably by preventive cultural techniques. Botanical pesticides prepared at farm from local plants, animals and microorganisms are allowed. Use of synthetic chemicals such as fungicides, insecticides, herbicides, synthetic growth regulators and dyes are prohibited. Use of genetically engineered organisms or products is prohibited.

Soil and Water Conservation

Soil and water resources should be handled in a sustainable manner to avoid erosion, salination, excessive and improper use of water and the pollution of surface and ground water. Cleaning of land by burning (*e.g.* slash and burn and straw burning) should be restricted. Clearing of primary forest for agriculture (jhuming or shifting cultivation) is strictly prohibited.

3) Collection of Non-cultivated Material of Plant Origin and Honey

Wild harvested products shall only be certified organic, if derived from a stable and sustainable growth environment and the harvesting shall not exceed the sustainable yield of the ecosystem and should not threaten the existence of plant or animal species. The collection area should not be exposed to prohibited substances and should be at an appropriate distance from conventional farming, human habitation, and places of pollution and contamination.

4) Pest and Disease Control

For pest management and control following measures shall be used in order of priority Preventive methods such as disruption, and elimination of habitat and access to facilities. Other methods of pest control are:

1) Mechanical, physical and biological methods
2) Permitted pesticidal substances as per the standards and
3) Other substances used in traps.

Irradiation is prohibited. Direct or indirect contact between organic products and prohibited substances (such as pesticides) should not be there.

5) Food Processing and Handling

Organic products shall be protected from co-mingling with nonorganic products, and shall be adequately identified through the whole process. Certification programme shall regulate the means and measures to be allowed or recommended for decontamination, clearing or disinfection of all facilities where organic products are kept, handled, processed or stored. Besides storage at ambient temperature, there must be cooling, freezing, drying and humidity regulation.

6) Packaging

Material used for packaging shall be eco-friendly. Unnecessary packaging material should be avoided. Recycling and reusable systems should be used. Packaging material should be biodegradable. Material used for packaging shall not contaminate the food.

7) Labeling

When the full standard requirements are met, the product can be sold as "Organic". On proper certification by certification agency "India Organic" logo can also be used on the product.

8) Storage and Transport

Products integrity should be maintained during storage and transportation of organic products. Organic products must be protected from co-mingling with non-organic products and must be protected all times from contact with the materials and substances not permitted for use in organic farming.

Grower Group Certification System

This system is based on the internal quality system and shall apply to producer groups, farmer's cooperatives, contract production and small scale processing units. The producers in the group must apply similar production systems and the farms should be in geographical proximity.

Constitution of Group

The group should have a legal status or constitution of the organization and shall be presented by an organizational chart.

Internal Quality System (IQS)

Group certification is based on the concept of an Internal Quality System comprising of the following:

- ☆ Implementation of the internal control system
- ☆ Internal standards
- ☆ Risk assessment

An external inspection and certification body should be identified for conducting annual inspection of the individual group/unit. The external inspection agency shall evaluate by checking the IQS documentation, staff qualifications and re-inspecting some farms.

Developing IQS

The following are the minimum requirements for setting up an IQS for grower groups:

- ☆ Development of Internal Control System (ICS)
- ☆ Identification of producer groups

- ☆ Creation of awareness about group certification
- ☆ Identification of qualified personnel for maintaining the internal control system
- ☆ Give necessary training in production and IQS development
- ☆ Preparation of IQS manual containing policies and procedures
- ☆ Implementation of the policies and procedures
- ☆ Review and improvement of the IQS document for maintaining a harmonized IQS

The system is operated through following workers:

1) Internal quality system manager (IQS Manager)
2) Internal inspectors
3) Approval manager/committee
4) Field officers
5) Purchase officers
6) Warehouse manager
7) Processing manager (in case of processing unit)

Internal Standards

The internal standards shall be prepared in local language by the IQS manager for the region of operations under the framework of NPOP standards. If the farmers are illiterate, the internal standards shall contain illustrations in the text for better understanding.

Conflict of Interest

The IQS personnel shall not have any conflict of interest that might hinder the work. All possible conflicts shall be declared in a written statement. In such cases, the IQS shall ensure that alternative solutions are found.

Scope of Certification

The certification shall be granted to the group with reference to the regulations/ standards adopted by the group.

Internal Inspections

At least two inspections of the group (one in growing season of each crop) shall be carried out by the internal inspector and will be documented. The inspection will be carried out in presence of the member or his representative and must include a visit of the whole farm, storage of inputs, harvested products, post harvest handling and animal husbandry. In case of non-compliance, the results will be reported to the IQS manager and all measures should be taken according to the internal sanction procedures.

External Inspections

The external Inspection and Certification Agency will re-inspect some of the farms for the evaluation of the grower group for efficient internal control system for compliance with the NPOP Standards. The sampling plan for inspection shall be based on the inspector's perception of risk.

Yield Estimates

Yields will be estimated for each crop for individual farmer in the group. This activity should be carried out especially during harvesting and should be counter-checked with the estimates during buying.

Certification Procedure in Brief

1) Application is made to the certification agency in the prescribed format with necessary farm and process details
2) Screening of application by certification agency and if necessary further details/clarification sought
3) Cost estimate comprising of certification charge, inspection charge, travel cost, reporting cost, laboratory charges *etc.* is sent for acceptance
4) Acceptance of cost by the grower/producer
5) Signing of agreement between grower/producer and certification agency
6) Certification agency seeks cropping/production/cultivation/processing plan and supply a copy of the standards to the grower/producer to follow
7) Certification agency raises an invoice and asks the producer to release 50 per cent of the certification cost in advance
8) Grower/producer pays the fee
9) Inspection schedule is worked out
10) Inspection is carried out at one or more than one occasion
11) If required unannounced inspection can also be done. In case of doubt the inspection team can also draw plant/soil/raw material/input/product sample for laboratory analysis.
12) Inspection report/(s) submitted to the certification committee
13) Certification agency asks for final payment
14) Final payment is made
15) Certification is granted
16) Grower/producer releases the stock for sale with Certification Mark (India Organic Logo)

Chapter 12

Labels, Labeling, and Market Information

The term, "organic" may only be used on labels and in labeling of raw or processed agricultural products, including ingredients, that have been produced and handled in accordance with the regulations. The term, "organic," may not be used in a product name to modify a nonorganic ingredient in the product. Products for export, produced and certified to foreign national organic standards or foreign contract buyer requirements, may be labeled in accordance with the organic labeling requirements of the receiving country or contract buyer.

Organic Product Composition

(a) **Products sold, labeled, or represented as "100 per cent organic":** A raw or processed agricultural product sold, labeled, or represented as "100 per cent organic" must contain (by weight or fluid volume, excluding water and salt) 100 per cent organically produced ingredients.

(b) **Products sold, labeled, or represented as "organic":** A raw or processed agricultural product sold, labeled, or represented as "organic" must contain (by weight or fluid volume, excluding water and salt) not less than 95 per cent organically produced raw or processed agricultural products.

(c) **Products sold, labeled, or represented as "made with organic (specified ingredients or food group(s)":** Multiingredient agricultural product sold, labeled, or represented as "made with organic (specified ingredients or food group(s))" must contain (by weight or fluid volume, excluding water and salt) at least 70 per cent organically produced ingredients which are produced and handled pursuant to requirements

(d) **Products with less than 70 per cent organically produced ingredients:** The organic ingredients in multiingredient agricultural product containing less than 70 per cent organically produced ingredients (by weight or fluid volume, excluding water and salt) must be produced and handled pursuant to requirements.

(e) **Livestock feed:** A raw or processed livestock feed product sold, labeled, or represented as "100 per cent organic" must contain (by weight or fluid volume, excluding water and salt) not less than 100 per cent organically produced raw or processed agricultural product. All products labeled as "100 per cent organic" or "organic" and all ingredients identified as "organic" in the ingredient statement of any product must not:

(i) Be produced using ionizing radiation

(ii) Be processed using sewage sludge

(iii) Be produced using nonorganic ingredients when organic ingredients are available; or

(iv) Include organic and nonorganic forms of the same ingredient.

Calculation of the Percentage of Organically Produced Ingredients

(a) The percentage of all organically produced ingredients in an agricultural product sold, labeled, or represented as "100 per cent organic," "organic," or "made with organic (specified ingredients or food group(s)," or that include organic ingredients must be calculated by:

(1) Dividing the total net weight (excluding water and salt) of combined organic ingredients at formulation by the total weight (excluding water and salt) of the finished product.

(2) Dividing the fluid volume of all organic ingredients (excluding water and salt) by the fluid volume of the finished product (excluding water and salt) if the product and ingredients are liquid. If the liquid product is identified on the principal display panel or information panel as being reconstituted from concentrates, the calculation should be made on the basis of single-strength concentrations of the ingredients and finished product.

(3) For products containing organically produced ingredients in both solid and liquid form, dividing the combined weight of the solid ingredients and the weight of the liquid ingredients (excluding water and salt) by the total weight (excluding water and salt) of the finished product.

(b) The percentage of all organically produced ingredients in an agricultural product must be rounded down to the nearest whole number.

(c) The percentage must be determined by the handler who affixes the label on the consumer package and verified by the certifying agent of the handler. The handler may use information provided by the certified operation in determining the percentage.

Marketing

Like all growers, many organic farmers find marketing is the hardest part of farming. One may not only need a superior product, but also need to get it into the hands of those who want it. Some lack the skills and creativity to find profitable outlets; some simply dislike dealing with the public, doing research, or addressing the other details essential to successful marketing. Further, transportation costs to and from market can seriously eat into profits. Therefore, farmers' markets have been booming in recent years, and these make an ideal outlet for organic growers to sell direct to consumers.

However, a positive action plan for sustainable organic farming should give financial support for farmers wishing to convert to an organic system (especially arable and horticultural producers) to supply the growing demand for organic produce and reduce the dependence on imports, as well as post-conversion organic stewardship payments for organic farmers in recognition of the health and environmental benefiting organic farming. The organic food industry is a booming business, and with the recent sale of natural-foods giant Whole Foods to Amazon, it's expected to grow even larger in the near future.

Fruits and Vegetables where the Organic Label matters Most

According to the Environmental Working Group, a nonprofit organization that analyzes the results of government pesticide testing, the following fruits and vegetables have the highest pesticide levels so are best to buy organic:

- ☆ Apples
- ☆ Cucumbers
- ☆ Potatoes
- ☆ Cherry Tomatoes
- ☆ Summer Squash
- ☆ Peaches
- ☆ Strawberries
- ☆ Sweet Bell Peppers
- ☆ Celery
- ☆ Grapes
- ☆ Kale/Collard Greens
- ☆ Nectarines (imported)
- ☆ Spinach
- ☆ Hot Peppers

Fruits and Vegetables need not to buy Organic

Known as the "Clean 15", these conventionally-grown fruits and vegetables are generally low in pesticides.

- ☆ Asparagus
- ☆ Mushrooms
- ☆ Sweet Corn
- ☆ Kiwi
- ☆ Onion
- ☆ Pineapple
- ☆ Sweet Potatoes
- ☆ Cantaloupe
- ☆ Avocado
- ☆ Cabbage
- ☆ Eggplant
- ☆ Mango
- ☆ Papaya
- ☆ Sweet Peas (frozen)
- ☆ Grapefruit

Organic Market in India

With the Indian organic food market anticipated to grow at a CAGR (Compound Annual Growth Rate) of over **25 per cen**t in the next three years, there is tremendous scope and opportunity for venturing in the business of organic food and organic farming. Global organic food market stood at $ 110.25 billion in 2016 and is projected to grow at a CAGR of 16.15per cent, in value terms, during 2017 – 2022, to reach $ 262.85 billion by 2022. Here are a few things to keep in mind if someone plans to do so:

1) Market Opportunity

It is always a good idea to do a market background check before entering into any industry. When it comes to the organic food market, one must be aware that with increasing demand, there are growing incidences of food adulteration, which are repeatedly reported across the globe. Consumer consciousness of natural, healthy and safe foods is on the rise, new food safety legislation is also working towards improving the safety standards of food and at the same time consumers are increasingly willing to pay for organic foods as their disposable incomes rise. All of this should be well considered before taking the plunge.

2) Awareness of Target Audience

One has to make sure to understand the target audience very well. This generation is educated and smart; thus they prefer to eat healthy and spend on healthy food than spend on medical treatments. The Indian consumer mindset is always **ready to adopt** the new. With health becoming a priority, MRP has taken a backseat when buying organic. With better spending power and with growing awareness, these products are not just becoming popular but are finding a sure place in the buying baskets. The market for organic food is growing everywhere and it is important to know that everyone is a potential buyer.

3) Growth Opportunity

This high-scale opportunity is driven by the fact that organic food consumption is not even 0.1 per cent of India's **$300 billion annual food consumption.** Likewise, India's share in the $100 billion global organic food market is less than 0.2 percent. The scale of the opportunity is driven by the increasing variety of organic foods hitting the market and the emergence of ecommerce as a channel in which to sell the product. This makes it a **good business model** to venture into or invest in. Another reason for investor interest is that the typical organic supply chain is more efficient compared to conventional agri-businesses as there are fewer intermediaries and there is in India's organic opportunity making it a lucrative market to venture into.

4) The Rise in Demand for Organic Food

As the digital literacy rate in India is growing, the demand for organic food in India will see good growth. People are becoming aware of the harmful effects of chemicals and fertilizers used during farming in India. There is a great rise in awareness of organic farming methods, making the food more healthy and nutritious. With free/low-cost access to the Internet, more and more people are getting to learn about the benefits of organic food. Therefore, the demand has gone up.

5) Higher Spending Capacity of the Middle Class

According to a research by the India Brand Equity Foundation, the spending capacity of the Indian middle class is anticipated to double by 2025. With the rise in disposable income and access to credit, the Indian middle class will have a high spending capacity on organic food. A growth in e-commerce sector has acted like a facilitator for the organic food industry to reach out to the potential consumers in Tier II and Tier III cities.

6) Rise in the Export of Indian Organic Food

According to the Agricultural and Processed Food Products Export Development Authority (APEDA), the demand for Indian organic food products is on the constant increase worldwide as India exported organic products worth \$515 million in the financial year 2017-18, from \$370 million in 2016-17. This trend will further accelerate in coming years.

7) Supply Chain Enhancement

One of the major issues in the organic food market is the mismanagement in the supply chain of organic fruits and vegetables. The main challenge is related to the cost and logistics which are included in the moving of locally or regionally produced organic produce. In the case of organic supplies, the time of reaching the fruits and vegetables from the time of the production to the end consumers matters the most due to its less shelf life. But since the demand of the organic products is rising, the need for solving the problems in the supply chain management is going to arise and the enhancement in the supply chain is expected in near future.

8) Government Regulations

The government of India is promoting Organic Farming in the country. Government of India has introduced Paramparagat Krishi Vikas Yojana (PKVY), under which 2 lakh hectares has been made suitable for organic farming thereby benefitting 5 lakh farmers. This step will further boost organic farming in India.

9) Challenges and Policy Changes

It is important to be aware of the challenges which will be faced and to be side by side with the policies and norms in this sector.

a) The government is subsidising farmers under the Participatory Guarantee System (PGS) for India, which is a self-certification process supported through the Pradhan Mantri Kaushal Vikas Yojana (PKVY) scheme, but these farmers are not allowed to export. Agricultural and Processed Food Products Export Development Authority (APEDA) has made it mandatory to have a third-party certification for exports. This is despite the fact that globally more than 100 developing countries recognise the PGS. Unless farmers under PGS India are allowed to export, they cannot earn the premium price. Therefore, ideally, farmers have to judicially decide where they want to sell the product—domestic market and/or export market.

b) There is need for a policy on input standardization.

c) Different varieties of crops are grown in different regions of the country, and they are faced with different issues related to pest infestation and soil quality. Hence, there is a need for more crop-specific and region-specific research and development on organic inputs.

d) Farmers need access to equipment such as netting and poly-houses to protect their crops against insects.

Present Status of Organic Market in India

Organic food in India is a niche category and the wave which is still to sweep across the nation. Growth in e-commerce sector is acting as facilitator for the companies as an appealing route to the potential consumers as channels margins in ecommerce are lower. There are dedicated as well as generalist e-commerce websites selling organic food. Companies are also opening up small kiosk at malls across major metropolitan cities. Organic beverages like tea have the highest value share followed by pulses and dairy products.

The major demands under the organic product category are for oil seeds, cereals and millets, sugar, fruit juice concentrates, tea, spices, pulses, dry fruits, medicinal plant products *etc.* Though USA, European Union member countries and Canada were the biggest buyers of organic products, many new countries like Israel, Vietnam, Mexico are also taking interest recently.

India is also negotiating with Canada, South Korea, Taiwan and Japan for equivalency with NPOP. In addition the traceability system, Tracenet established by APEDA for certification and export of organic products also helps India to maintain the credibility and traceability of certified products.

Even though India has the highest number of farmers currently engaged in organic farming (about 8,35,000) the country accounts for less than 1 per cent the global organic market price. In terms of total area under organic cultivation, India ranks at 9th position with 1.49 million hectares. Australia, which tops the list, has more than 27 million hectares under organic farming, it said.

Organic Packaged Food Market

Organic packaged food and beverages is an emerging niche market in India and its primary consumers are high-income urbanites. The total market size for organic packaged food in India in 2016 was INR533 million, growing at 17 per cent over 2015, and is expected to reach INR871 million by 2021. India's exports of organic products have been increased by 17 per cent between 2015-16 and 2016-17. In India, the majority of the demand comes from tier 1 cities16. Companies are witnessing notable growth as demand from metro cities increase with the entry of several new players in the organic food market such as Conscious Foods, Sresta,

Eco Farms, Organic India, Navdanya and Morarka Organic Foods to name a few. In addition to the growing domestic market, India is the second largest exporter of organic products in Asia after China. The increasing export market coupled with the Government's support is making organic cultivation in India highly successful.

Indian organic market is retailer-driven and aggregator-powered and there is a need to make it consumer-driven and farmer-powered. 7 e-commerce retail chains like Big Basket and Godrej Nature Fresh are also attempting to capitalize on this growth spurt, whereas food companies like Elite Food have planned to launch their organic foods including sugar, spices and pulses. Established players in the organic market segment are optimistic about their growth trajectory.

The organic markets in India are largely spread across the food and beverages, health and wellness, beauty and personal care and textile industries. The highest growth is observed in the organic food segment, followed by textile, beauty and personal care. The current Indian domestic market is estimated at Rs 40,000 million which is likely to increase by Rs 100,000 million — Rs 120,000 million by 2020 with a similar increase in exports.

Organic Processed Food

With the rapid growth in organic food consumption, the market for processed organic products also increases. The term processing implies activities such as cleaning, cutting, heating, canning, and freezing as well as the addition of preservatives, flavors, or other substances approved for use in food products, such as preservatives, salt, sugars, and fats.

Availability of organic alternatives to conventional food is another important determinant of organic food consumption. Consumers want organic products to be easily available in all shops, and there is also an increasing market for home delivery of boxes of organic fruit and vegetables. Consumers are prepared to pay more for products that involve less processing and that are grown free of pesticides and genetically modified organisms, perhaps by local farmers considered worthy of support. Insufficient availability, convenience concerns, and high prices for organic food are identified as factors that limit the consumption of organic food. Therefore, agricultural products being sold as organic may be transported only in packaging or containers closed in a manner which would prevent substitution of the content.

Labeling

Labeling and advertising may refer to organic production methods only where they make it clear that the information relates to a method of agricultural production. The product concerned must comply with the provisions of Regulation. Moreover, the operator must be subject to the inspection measures laid down in the Regulation, and the name and/or code number of the inspection authority must be indicated. The rules on indications referring to organic production methods stipulate the minimum percentage of agricultural ingredients that must be of organic origin.

Labeling and advertising of a food product may bear indications referring to organic production methods in the sales description only where at least 95 per cent of the ingredients of agricultural origin are organic. Food products may thus contain up to 5 per cent of ingredients produced by conventional methods as long as those ingredients are not available (*e.g.* exotic fruit) or in very short supply on the Community organic market.

Where the ingredients of organic origin represent less than 70 per cent of the content of a product, the labeling and advertising may not bear any reference to organic production methods.

Advertising

Regulation also provides for certain conditions to govern advertising of organic products. The purpose of these rules is to guarantee that advertising will not compromise the effects of the rules intended to ensure transparent labeling.

Inspection

In view of the importance of maintaining the credibility of the organic farming sector, the regulation introduces a number of rules on the inspection of operators. All operators who, as part of a business activity, produce, prepare or import agricultural products or foodstuffs obtained by organic production methods are subject to a special inspection scheme. The inspection body must inspect each holding at least once a year, and may also visit farms without prior warning. Inspectors must check on compliance with the rules, and may take samples so as to test for the presence of unauthorized products.

Importers of Organic Products

The rules for inspection of importers are also intended to supervise the movements of each individual consignment of products imported from third countries, in particular by requiring full identification of products (quantity, type, origin). The inspection body must be provided with information on the transport and consignees of products.

Sanctions for Failure to Comply with Community Rules

Stricter penalties may be imposed where a manifest infringement or an infringement with prolonged effects is found; in that case, the operator may be prohibited from producing or marketing organic products for a period to be determined by the inspection body.

Better Understanding of the Economics of Organic Farming and the Potential for Environmental Markets

Organic farming builds community and provides societal benefits including the jobs created by thriving farms, vibrant farmers markets and food co- ops, and prosperous processors. The local and organic food movements have emerged as powerful forces in the marketplace and in communities. These issues are ripe for in-depth investigation by economists, social scientists, rural sociologists, and others. Environmental markets depend on placing an economic value on credits and performance, a difficult interdisciplinary task requiring ecologists and economists to collaborate. Valuation rules should recognize the multiple functions of natural resources and how organic farming systems provide multiple services by managing the land in a sustainable manner. Ideally, organic interests must be engaged in the creation of the environmental market in order to clearly articulate how and why certified organic farms are high performers. Thus far, valuation systems are evolving and are the current focus of much research that may bear fruit in the near term.

Chapter 13

Shortcomings of Organic Farming

It is quite natural that a change in the system of agriculture in a country of more than a billion people should be a well thought out process, which requires utmost care and caution. The world is taking up a healthier turn for good as people seem to accept the inclusion of organic food as a part of their daily life. Yet, the rate of adoption has been slow in the Indian market. Indian population which represents almost 20 per cent of the world's population consumes less than 1 per cent of world's total organic produce, despite having the largest organic farmland. Its growth has been sluggish, with people holding preconceived notions about the segment. Here are a few hindrances that this segment is facing.

1) Lack of Awareness of Consumers

The organic food industry is a fledgling industry where its consumers mostly belong to the "fad diet" category, with a great lacking in awareness. Working on a word by mouth principle, core awareness for the category is negligible, with many people opting for these options as they seem fancy. This deficiency in knowledge creates consumers that glamorize the organic consumption, creating temporary demand. This, in turn, affects the sustainable growth of the whole industry. The concern for safe food is still far away.

2) Lack of Awareness of Producers

It is a fact that many farmers in the country have only vague ideas about organic farming and its advantages as against the conventional farming methods. Use of bio-fertilizers and bio pesticides requires awareness and willingness on the part of the farming community. Knowledge about the availability and usefulness of supplementary nutrients to enrich the soil is also vital to increase productivity.

Farmers lack of knowledge of compost making using the modern techniques and also its application. The maximum they do is making a pit and fill it with small quantities of wastes. Often the pit is flooded with rainwater and result is the top of the compost remains under composted, the bottom becomes like a hard cake. Proper training to the farmers will be necessary to make vermi-compost on the modern lines.

Attention on the application of composts/organic manure is also lacking. The organic matter is spread during the months when the right moisture level is absent on the soil. The whole manure turns into wastes in the process. The required operation is of course labor intensive and costly, but it is necessary to obtain the desired results.

3) Output Marketing Problems

It is found that before the beginning of the cultivation of organic crops, their marketability and that too at a premium price over the conventional produce has to be assured. Inability to obtain a premium price, at least during the period required to achieve the productivity levels of the conventional crop will be a setback.

4) Shortage of Bio-mass

Many experts and well informed farmers are not sure whether all the nutrients with the required quantities can be made available by the organic materials. Even if this problem can be surmounted, they are of the view that the available organic matter is not simply enough to meet the requirements.

The crop residues useful to prepare vermi-compost are removed after harvest from the farms and they are used as fodder and fuel. Even if some are left out on the farms termites, *etc.* destroy them. Experiments have shown that the crop residues ploughed back into soil will increase productivity and a better alternative is conversion into compost.

The small and marginal cultivators have difficulties in getting the organic manures compared to the chemical fertilizers, which can be bought easily, of course if they have the financial ability. But they have to either produce the organic manures by utilizing the bio-mass they have or they have to be collected from the locality with a minimum effort and cost. Increasing pressure of population and the disappearance of the common lands including the wastes and government lands make the task difficult.

5) Inadequate Supporting Infrastructure

In spite of the adoption of the NPOP during 2000, the state governments are yet to formulate policies and a credible mechanism to implement them. There are only four agencies for accreditation and their expertise is limited to fruits and vegetables, tea, coffee and spices. The certifying agencies are inadequate, the recognized green markets are non-existent, the trade channels are yet to be formed and the infrastructure facilities for verification leading to certification of the farms are inadequate.

6) High Input Costs

The small and marginal farmers in India have been practicing a sort of organic farming in the form of the traditional farming system. They use local or own farm renewable resources and carry on the agricultural practices in an ecologically friendly environment. However, now the costs of the organic inputs are higher than those of industrially produced chemical fertilizers and pesticides including other inputs used in the conventional farming system. The groundnut cake, neem seed and cake, vermi-compost, silt, cow dung, other manures, *etc.* applied as organic manure are increasingly becoming costly making them unaffordable to the small cultivators.

7) Marketing Problems of Organic Inputs

Bio-fertilizers and bio-pesticides are yet to become popular in the country. There is a lack of marketing and distribution network for them because the retailers are not interested to deal in these products, as the demand is low. The erratic supplies and the low level of awareness of the cultivators also add to the problem. Higher margins of profit for chemical fertilizers and pesticides for retailing, heavy advertisement campaigns by the manufacturers and dealers are other major problems affecting the markets for organic inputs in India.

8) Absence of an Appropriate Agriculture Policy

Promotion of organic agriculture both for export and domestic consumption, the requirements of food security for millions of the poor, national self-sufficiency in food production, product and input supplies, *etc.* are vital issues which will have to be dealt with in an appropriate agriculture policy of India. These are serious issues the solution for which hard and consistent efforts along with a national consensus will be essential to go forward. Formulation of an appropriate agriculture policy taking care of these complexities is essential to promote organic agriculture in a big way.

9) Lack of Financial Support

The developing countries like India have to design a plethora of national and regional standards in attune with those of the developed countries. The adoption and maintenance of such a regulatory framework and its implementation will be costly. The cost of certification, a major component of which is the periodical inspections carried out by the certifying agencies, which have freedom to fix the timings, type and number of such inspections appears to be burdensome for the small and marginal farmers. The fees charged by the international agencies working in India before the NPOP were prohibitive and that was a reason for the weak response to organic agriculture even among the large farms in the country. No financial support as being provided in advanced countries like Germany is available in India. Supports for the marketing of the organic products are also not forthcoming neither from the State nor from the Union governments, even the financial assistance extended to the conventional farming methods are absent for the promotion of organic farming.

10) Lack of Quality Standards for Biomanures

The need for fixing standards and quality parameters for bio-fertilizers and biomanures has arisen with the increasing popularity of organic farming in the country. There are a large number of brands of organic manures, claiming the high levels of natural nutrients and essential elements. But most farmers are not aware of the pitfalls of using the commercially available biomanure products. While the concept of organic farming itself lays great stress on the manures produced on the farm and the farmers' household, many of the branded products available in the market may not be really organic. Elements of chemicals slipping into the manures through faulty production methods could make the product not certifiable as organic.

11) Improper Accounting Method

An understanding of the real costs of erosion of soil and human health, the loss of welfare of both humans and other living things and the computation of these costs are necessary to evaluate the benefits of organic farming. These costs will have to be integrated to a plan for the implementation of organic agriculture.

12) Consumer Preference for Low Price Product

India is a price sensitive nation, with great disparity in the earning power. Organic food, due to various factors is steeper on price. Unaware consumers are wary of spending that much on organic food. Huge difference between the pricing of organic and the conventional product price has created disfavour towards organic food.

13) Pay Extra Percentage to the Retailer

Retailers prefer conventional over the organic for the quick movement from the shelf and hence for bringing in a greater margin per sq. inch of the shelf space, because organic material replenishment cycle is much more as compared to the conventional product. Retailers demand a good premium in order to create a shelf space to showcase the organic product. There is a push from the manufacturer and the brand than the pull from the market because the whole of the industry is new, adding a scope for the retailer to further assume an upper-hand. So the brands have to give a lot more percentage to the retailer than the conventional brand to get the shelf space. That makes the whole organic thing more expensive than it should be.

14) Expensive Storage Method

Apart from pricing, a hike of expenses is noted in terms of storage and preservation. Very expensive storage method is being used as chemical input is not utilized to store these products for long-term. Conventional farmers enlist the use of chemicals to reduce their loss of crops, while organic farmers have to remain restricted to limited natural methods. Without the perquisite chemical preservatives added to conventional foods, organic foods face a shorter storage time and shelf life and hence require a greater level of maintenance.

15) Supply-Demand Disparity

The basket of the organic production is not balanced. What is in more demand is lacking in supply *e.g.* fresh fruits and vegetables while the supply of some of the grains outgrows the demand. Hence, there is the disparity in terms of what is needed by the consumers versus what is provided by the producers. This gap needs to be filled by local production in clusters.

16) Higher Labour Input in Organic Farming

Organic farming requires a greater volume of labour input. The labour input measured in terms either of hours of work or of full-time jobs is usually greater on organic than on equivalent conventional farms. The main reason why organic farming requires more labour is to carry out manual and mechanical tasks essential to growing. The preparation of products for sale on the farm or on the market also involves more labour on organic holdings.

17) Lower Productivity

An organic farm cannot produce as much yield as a conventional or industrialized farm. A 2008 survey and study conducted by the UN Environmental Program concluded that organic methods of farming result in small yields even in developing areas, compared to conventional farming techniques. Though this point is debatable as the productivity and soil quality of an industrialized farm decreases rapidly over the years.

18) Requires Skill

An organic farmer requires greater understanding of his crop and needs to keep a close watch on his crops as there are no quick fixes involved, like pesticides or chemical fertilizers. Sometimes it can be hard to meet all the strenuous requirements and the experience to carry out organic farming.

19) Time-Consuming

Significant amounts of time and energy are required to execute the detailed methods and techniques that are required for a farm to be called an organic farm. Failure to comply with any of these requirements could result in loss of certification, which the farmer will not be able to regain in up to three years and it can be more time-consuming. Organic farming increases soil fertility by way of compost, and organic fertilizers and mulch. Organic fertilizers tend to be slow-release. As with control by botanicals, horticultural oils, and insecticidal soaps, organic fertilizers may need several applications before the desired results are brought about.

20) Political and Social Factors

Agriculture in India is subject to political interventions with the objectives of dispensing favours for electoral benefits. Subsidies and other supports from both the Central and state governments, government controlled prices of inputs like chemical fertilizers, the public sector units' dominant role in the production of fertilizers, government support/floor prices for many agricultural products, supply of inputs like power and water either free of cost or at a subsidized rate, *etc.* are the tools often used to achieve political objectives. Any movement for the promotion of organic farming in India will have to counter opposition from the sections who benefit from such policies in the conventional farming system. The political system in a democracy like India is likely to evade the formulation of policies, which affect the interests of the voting blocks unless there are more powerful counter forces demanding changes.

As a farmer converts his/her land from conventional chemical-based farming to organic farming, there is a risk of loss in yield due to the withdrawal of chemical inputs and high-yielding varieties of seeds. In India, there is **no subsidy** to compensate for the yield loss during the conversion period unlike UK. Further, a majority of the government budget and subsidies are targeted towards chemical-based inputs and, in many states, less than two per cent of the budget is allocated to organic farming.

Cons of Organic Foods

Sometimes, the belief is expressed that organic fruits or vegetables are "full of healthy stuff", while conventional food is "empty". There is no scientific base for this belief. If large differences existed under present farming practices, they would have been found by now. While it may seem like the healthier alternative to adopt a strictly organic food diet, there are some downsides to it:

- Organic foods are significantly more expensive than conventional foods. On average, organic food is 47 per cent more expensive than regular food. A study in the Proceedings of the National Academy of Sciences found that it's only 5-7 per cent more expensive, so the price difference is not really justified; it also means that organic farming has become more profitable than regular farming. In other words, from a price point of view, organic has become a synonym for luxury.
- One of the major challenges faced by consumers who buy organic products by paying so much premium price is that whether they can trust the product to be authentic or not. Organic products are currently sold in the country completely on the basis of trust and consumer has almost no way to check the authenticity of the claim.
- Not all conventionally grown foods contain residue nor are organics always the more virtuous choice. As the skin can protect the fruit or vegetable from any pesticide exposure so when the outside can be peeled away, it may not be worth spending the extra cash for organic.
- Despite the widespread conviction held by the public that organic food is 'healthier' than foods produced using conventional farming, evidence to support this perception is difficult to identify.
- It is not easy to have a wide selection of organic foods in all food stores have.
- Organic foods have a shorter shelf life compared to conventional foods so they must be consumed relatively soon after purchase.

- Since there are very little to no pesticides and herbicides used on organic foods, consumption of organic foods have been linked to an increased risk of *E. coli* bacteria in the digestive system.
- In 2012, a massive meta-analysis was published by Stanford team which analyzed 240 studies: 17 comparing populations consuming organic and conventional diets, and 223 studies that compared either the nutrient levels or the bacterial, fungal or pesticide contamination of various products (fruits, vegetables, grains, meats, milk, poultry, and eggs) grown organically and conventionally. They report little significant difference in health benefits between organic and conventional foods, as well as no consistent differences in the vitamin content of organic products. In fact, only one nutrient (again, phosphorous) was significantly higher in organic versus conventionally grown produce. Protein and fat content were also similar, although a significant difference was reported in organic milk, which contained higher levels of omega-3 fatty acids.

Therefore, although there is a common perception, perhaps based on price alone, that organic foods are better for you than non-organic ones, it remains an open question as to the health benefits.

Chapter 14

Organic Farming in India

Development of agricultural practices ensuring food sufficiency is a basic human requirement, a prerequisite for satisfactory social conditions and a necessity for civilizations to flourish. India has a long history of successful biodiversity friendly agriculture, with introduced chemical agriculture practices. Indian agriculture dates back more than 4000 years, maintaining soil fertility over this period, and therefore being self-sustaining. During green revolution in India, productivity increased due to advancements in technology, irrigation, electrification, research, support prices and subsidies. High yielding varieties were introduced and supported by Government, which increased yields with use of fertilizers.

As India is second most populous country in the world; it has a large impact on environment, especially now, under a rapid economic growth. The growth of population and the increase in income will lead to a rise in demand for food grains as also for the agricultural raw materials for industry in the future. Food production and therefore agriculture has a special importance in India.

About 80 per cent of geographical area of India is under agriculture, forestry, pasture and biomass production. The agriculture and allied sectors in India provide employment to 65 per cent of the workers and accounts for 30 per cent of the national income. The area under cultivation, obviously, cannot be increased and the present 140 million hectares will have to meet the future increases in such demands. There is a strong reason for even a decline in the cultivated area because of the urbanization and industrialization, which in turn will exert much pressure on the existing, cropped area. Because of expanding urban and agricultural areas, area under natural pasture is declining causing cattle death. Because of insufficient animal feed, essential nutrients in excreta is less. Poor population of India use animal manure for domestic fuel that way limiting its availability for soil amendment. Tillage and mulch farming have little availability as most of the crop residues are used as fodder and fuel. Farmers in India face high prices of costs of seeds and fertilizers and cheap imports from heavily subsidized foreign countries Agriculture has become cash based individual enterprise requiring high investments in modern inputs and wage labors. Therefore farmers rely on credits. Without remunerative prices and with possible crop fails, farmers' economics may be easily destroyed. Uncertainty of yield can cause big problems, and even when yield is good, fluctuation of prices can have great impact. Moreover, agriculture becomes no longer object of attraction or site of investment.

Under these circumstances, one of concerns faced by country is food insecurity, which should be addressed in part by agriculture by effective food production. The progress of organic agriculture in India is very slow. According to FAO (2006), global food production needs to increase by 70 per cent to feed a population of 8.9 billion in 2050 – assuming the continuation of current dietary trends and patterns of global production and trade.

India has the potential to become a major organic producing country given the international demand for our farm products, different agro-climatic regions for the cultivation of a number of crops, the size of the domestic market and above all the long tradition of environment friendly farming and living. The need for organic farming in India arises from the unsustainability of agriculture production and the damage caused to ecology through the conventional farming practices.

Organic agriculture has been one of the fastest growing sectors in the global food industry for over two decades. India is bestowed with lot of potential to produce all varieties of organic products due to its agro-climatic regions. In several parts of the country, the inherited tradition of organic farming is an added advantage. This holds promise for the organic producers to tap the market which is growing steadily in the domestic market related to the export market. A country like India can enjoy a number of benefits from the adoption of organic farming. The price premiums for the products, conservation of the natural resources in terms of improved soil fertility and water quality, prevention of soil erosion, preservation of natural and agro-biodiversity are major benefits. Economic and social benefits like generation of rural employment, lower urban migration, improved household nutrition, local food security and reduced dependence on external inputs will be large gains in the Indian conditions. The protection of environment and the consequent increase in the quality of human life will be other contributions of organic farming.

The increasing demand for organic products is creating opportunities, and at the same time, majority of small farm holders are dependent on government incentives to meet the cost of input. Local farmers in India may be resistant to change their conventional practices, into organic ones. The main focus is on attitude towards organic farming of Indian farmers, as India has one of the richest biodiversity, and has a rapid development. Here, some factors have been discussed below which may influence farmers' decisions towards organic (sustainable) farming:

Investments

Factors such as uncertain markets and low prices for sustainable production, investments into new equipment, greater labor needs, seasonal labor, more facilities for transport and storage, changing the management style may create significant barriers for adopting sustainable practices. There is a fear among farmers for increased losses to insects, weeds, and their potential income will not justify those loses. They perceive that it is long term investment without immediate benefit. This is especially important in developing countries, as poverty typically impart a short run perspective to farm decision making.

To be certified organic during the conversion period of few years has to be grown organic but sold at conventional price. When failed to market the organic product, it must be sold at conventional prices and certification itself creates extra cost. Farmers who don't have significant problems with conventional methods may be unwilling to take risks for adoption, or the ones who would be willing may not have sufficient funds for it.

Markets

Farmers participate in the markets as buyers and sellers. They buy input from suppliers (seeds, fertilizer, machinery *etc.*) and sell their output. Therefore, direction of markets for farm inputs and products influence farmer production decisions. High prices and low variety of input may both decrease and stimulate willingness to change practices into more sustainable ones. Consumer demand and willingness to pay price premiums for sustainable produce are essential, as it provides opportunity for organic farmers to supplement their incomes.

Policy

Government policy is another factor that can do much to influence farmers' decisions. Macro level policies are important although they influence sustainable agriculture adoption incidentally. Exchange rate reforms, trade reforms, domestic agriculture price policies, labor market policies, rural public education all have significant impact on incentives or disincentives for sustainable agriculture adoption.

Biophysical Characteristics

Unlike many conventional farming practices, organic agriculture often is location-specific, meaning that they are not broadly generalizable and easily scalable. To adopt sustainable practices they have to be compatible with biophysical farm characteristics – geographical area and climate and the specifics of the farm itself, for example weeds, soil type, terrain, erosion potential, area cultivated, yield, slope of the land, *etc.*

Compatibility

Farmers also shown to be resistant to adopt agricultural practices that doesn't comply with current production strategies. Farming system normally is not easy to change and has been developed in years through trial and error, therefore new practices that are not compatible with current ones are less likely to be adopted.

Land Tenure

Land tenure is another factor influencing adoption. Renting the land results that the farmers don't have a complete control over the land they farm. This often has negative effect on adoption of organic practices. This can be due to land owner unwillingness for farmer to adopt sustainable practices which may alter appearance of land, and farmers' unwillingness to invest long-term in land that he does not own.

Knowledge and Information Awareness

Knowledge and information is vital component when adopting new farming technology. Farmers first have to be aware of organic agricultural practice, its implementation, potential benefits and drawbacks. Legislation, markets and other contextual conditions must be clear to dispel misconceptions, understand consequences and future prospects. It is believed that farmers in many developing countries lack technical information about farming practices, production and marketing methods, such as choosing products to grow, identifying different markets and distribution channels, competition strategies and market access. To gain necessary knowledge, to reduce risk by breaking free from misconceptions and understanding the aspects of sustainable agriculture, information and training must be made available and easy accessible.

Complexity

Agricultural systems for organic farming may be very complex. This complexity must come with knowledge about it so it could be implemented in full capacity, otherwise farmers will not be attracted to changes that require such intellectual investment.

Managerial Skills

Better managerial skills are required for sustainable farming practices. They are intensive and require commitment to constant learning. Inadequate managerial skill is one of the reasons for farmers being unable to adopt.

Age

Older farmers are less likely to consider sustainable practices. Older farmers may be reluctant as they don't have years ahead of them to see all the benefits from sustainable farming.

Farm Workers

Organic growers see reluctance among the workers to adapt to new practices for several reasons. Under conventional farming workers are exposed to chemicals which can be severe to their health, but according to farmers they do not realize the connection, when the effect after the cause is not immediate and when it is not obvious they do not understand the harm. This gets especially complicated in farms where the owner is not directly involved in the work and does not interact directly to his workers. If the owner decides to go organic must convince and change the mind-set of the staff, which in turn have to convince workers.

Marketing Skills

Converting from conventional to natural farming, one has to have or develop marketing skills. If one farmer being organic for some time can't market the product sufficiently; therefore not getting price premium cannot sustain the profits and consequently have to give up organic

The organic food market in the world has grown rapidly in the past decade. International trade in organic foods showed an annual growth rate of about 20-22 per cent during this period. Many retail chains and supermarkets in advanced countries are accorded with 'green status' to sell organic foods. The organic food processing industry is considered nature friendly and thus encouraged. The US food market tops, followed by Germany and Japan and France. Moreover, the organic foods command higher prices which often range from 10 to 100 per cent more than the conventional ones.

Organic farming movements have spread to Asia and Africa too. There are very large organizations promoting the organic farming movement in European countries, America and Australia *etc.* These organizations, for example, the International Federation of Organic Agriculture Movements (IFOAM) and Greenpeace have studied the problems of the chemical farming methods and compared the benefits accruing to the organic faming with the former.

International Federation of Organic Agriculture Movements (IFOAM)

IFOAM was founded in France in 1972. It spearheads and coordinates organic farming efforts the world over by promoting organic agriculture as an environment friendly and sustaining method. It focuses on organic farming by highlighting the minimum pollution and low use of non-renewable natural resources through this method. It has about 600 organizational members spread over about 120 countries including India. IFOAM undertakes a wide range of activities related to organic farming such as exchanging knowledge and thoughts among its members; representation of the movement in governmental, administrative and policy making forums in the national and international arena; updating of production, processing and trading standards; formulation and coordination of research projects; and holding of international conferences and seminars.

The Food and Agriculture Organisation (FAO)

FAO provides support to organic farming in the member countries. It also attempts the harmonization of national organic standards, which is absolutely essential to increase international trade in organic products. The FAO has, in association with the World Health Organisation (WHO), evolved the Codex Alimentarius for organic products.

With increasing demand of organic product in global market, several organizations in India showed keen interest in organic farming in last few years. The first conference of NGOs on organic farming in India was organized by the Association for Propagation of Indigenous Genetic Resources (APIGR) in October 1984 at Wardha. Several other meetings on organic farming were held at different places in the country towards the end of 1980s. Here, mention must be made of the Bordi Conference in Maharashtra, the state which was the focal point for the organic farming movement in India. The Rajasthan College of Agriculture with the support of the state government organized a meeting on organic agriculture in 1992. The United Planters' Association of South India (UPASI) organized two national level conferences on organic farming in 1993 and 1995. ARISE (Agricultural Renewal in India for a Sustainable Environment) is a major organization in the country engaged in the promotion of organic farming. ARISE was founded in 1995 at a national conference of organic farming held at Auroville. ARISE comprises of a supporting network of regional groups aiming at sustainable environment by protecting bio-diversity and promoting organic agricultural practices. The selection of Auroville for the conference was apt as it housed the Arabindo Ashram and the pioneering work under its auspices on building technology, alternative energy research, wasteland development, afforestation and organic agriculture.

By 1980, three groups of Indians had taken to organic farming. The first one consisted of urban educated technocrats for peripheral interest, which did not last long. Educated farmers consisted of the second group whose farming practices were based on scientific knowledge. The third group practiced organic farming through trial and error. The successful organic farmers in India are those who have access to sufficient natural resources like, water and other organic inputs mostly on their own farms. These farms produce crops like sugarcane, areca, cocoa, coconut, pepper and spices. Many of them have shown that switch over to organic farming do not affect yields and income and more importantly, knowledge/expertise is available for successful adoption of organic farming in the country.

The organic agriculture movement in India received inspiration and assistance from IFOAM which has about 600 organizational members from 120 countries. The International Federation of Organic Agriculture Movements (IFOAM) estimates that an area of about 41,000 hectares in India is under organic farming representing about 0.17 per cent of the world organic acreage. It also reveals that the percentage of organic area to the total cultivated area comes to only about 0.03. Currently, India ranks 33rd in terms of total land under organic cultivation and 88th position for agriculture land under organic crops to total farming area in the World.

National Program for Organic Production (NPOP)

Ministry of Commerce, Govt of India, has implemented the National Programme for Organic Production (NPOP) since 2001. The objectives of NPOP are as under:

1) To provide the means of evaluation of certification program for organic agriculture and products (including wild harvest, aquaculture, livestock products) as per the approved criteria.
2) To accredit certification programs of Certification Bodies seeking accreditation.
3) To facilitate certification of organic products in conformity with the prescribed standards.
4) To facilitate certification of organic products in conformity with the importing countries organic standards as per equivalence agreement between the two countries or as per importing country requirements.
5) To encourage the development of organic farming and organic processing.

The National Program for Organic Production (hereinafter referred to as 'NPOP') provides for Standards for organic production, systems, criteria and procedure for accreditation of Certification Bodies, the National (India Organic) Logo and the regulations governing its use. The standards and procedures have been formulated in harmony with other International Standards regulating import and export of organic products. This document also proposes to provide an institutional mechanism for the implementation of National Standards for Organic Production (hereinafter referred to as 'NSOP'). The NPOP is administered, monitored and implemented for the benefit of farmers, processors, traders and consumers. It envisages a three tiered organization under the overall guidance of the Union Government with the Department of Commerce, Ministry of Commerce and Industry as the nodal agency. Policy making and declaration of the standards for organic products, recognition of organic standards of other nations, efforts to get our standards recognized by others and coordination with other arms of the government for the successful management of the organic agriculture are the major functions entrusted to the ministry.

The Certification Bodies accredited under NPOP conduct inspection before certifying the operators as organic. After inspection and compliance to the requirement by the certified operators, certification decision is taken by the Certification Body and Scope Certificate is issued having a validity of one year. Organic products are traded only through issuance of Transaction Certificates (TCs) wherein the product identity and traceability is maintained. The TCs are also issued only by the accredited Certification Bodies. All the handlers of the product in the supply chain are required to be in the certification system under NPOP and need to obtain TC whenever the product is traded.

The NPOP standards for production and accreditation system have been recognized by European Commission and Switzerland for unprocessed plant products as equivalent to their country standards. Similarly, USDA has recognized NPOP conformity assessment procedures of accreditation as equivalent to that of US. With these recognitions, Indian organic products duly certified by the accredited certification bodies of India are accepted by the importing countries.

The agencies accredited are the Agricultural and Processed Food Products Export Development Authority (APEDA), Coffee Board, Tea Board and the Spices Board. The regulations cover exports, imports and the domestic trade of the organic products. But the government regulations are applicable to only the exports. So, an organic farm product can be exported only if it is certified by a certification body accredited for the purpose. The categories of products covered under accreditation are organic crop production, organic animal production, organic processing operations, wild products and forestry.

Government support is available for awareness creation, technology transfer, support for establishment of compost, biofertilizers, biopesticides and vermicompost production units. Financial assistance is being provided for capacity building for facilitation of certification and other. A national level steering committee is functioning as the apex advisory body for assisting the government to promote organic farming in the country. This body consists of representative's from the

Ministries of Agriculture, Food Processing Industries, Forests and Environment, Science and Technology, Rural Development and Commerce.

Organic Farming Association of India (OFAI)

The Organic Farming Association of India (OFAI) is umbrella for organic farmers in the country. It was set up by the senior most members of India's organic farming community in 2002. It was set up to promote organic farming, lobby with government agencies and departments to pay more attention to sustainable agriculture, and assist farmers using chemicals and pesticides to convert successfully to organic farming methods. Although organic agriculture is being promoted by government, there is no consistency in governmental programs in general.

AOFG India (Agriculture and Organic Farming Group India)

It is an NGO (Registered Public Charitable Trust U/s. Act 1882) working in the rain fed and hill slope farming areas of India since 2001. AOFG India started with a network of 35 NGOs geographically spread over North East India, Andhra Pradesh, Maharashtra, Uttrakhand, Himanchal Pradesh and Kerala and continues to support them in project development and implementation. The focused program is to benefit the rural communities in capacity building, human resources development and organic agriculture in rural development, crop value addition, marketing and income generation. Technical training and its application have taken centre stage in all the promotional activities. One of the key areas of activity of the NGOs is the promotion of organic farming, production, organic certification and marketing. Capacity building, quality extension activities are focal to farmer centered activities. AOFG India is working in close cooperation with Planning Commission of India, development ministries of Government of India, Small Farmers Agri Business Consortium, Agriculture Finance Corporation of India and financial institutions. AOFG India and Farmer Limiteds are working with Fresh Food Technology India, Fresh Food Technology Netherlands and Van Amerongem, Tiel the Netherlands in setting up processing facilities and specialized godown facilities of fumigation and vacuum packing facilities. All processing, storage and packing facilities are based on Oxygen and Carbon dioxide. No synthetic chemicals are used.

Organic Farming Status in India

As per the available statistics, India's rank in terms of World's Organic Agricultural land was 9th among 178 countries that actively practice organic agriculture and in terms of total number of producers was 1st (FIBL and IFOAM Year Book 2018). At present, the country is home to more than 835,000 organic producers, 699 processors, 669 exporters and 1.49 million ha area under organic cultivation. As on 31st March 2018, total area under organic certification process (registered under National Programme for Organic Production) is 3.56 million Hectare (2017-18). This includes 1.78 million ha (50 per cent) cultivable area and another 1.78 million Hectare (50 per cent) for wild harvest collection Among the states, Madhya Pradesh has the largest area under organic certification (4.62 lakh ha) followed by, Maharashtra (1.98 lakh ha) Rajasthan (1.55 lakh ha), Telangana

(1.04 lakh ha), Odisha (0.96 lakh ha), Karnataka (0.94 lakh ha), Gujarat (0.77 lakh ha) and Sikkim (0.76 lakh ha). These states had a combined share of 90 per cent of the area under organic certification in 2015-16.

During 2016, Sikkim has achieved a remarkable distinction of converting its entire cultivable land (more than 76000 ha) under organic certification. India produced around 1.70 million MT (2017-18) of certified organic products which includes all varieties of food products namely Oil Seeds, Sugar cane, Cereals and Millets, Cotton, Pulses, Medicinal Plants, Tea, Fruits, Spices, Dry Fruits, Vegetables, Coffee *etc.* The production is not limited to the edible sector but also produces organic cotton fiber, functional food products *etc.* Among different states Madhya Pradesh is the largest producer followed by Maharashtra, Karnataka, Uttar Pradesh and Rajasthan. In terms of commodities, Oil seeds are the single largest category followed by Sugar crops, Cereals and Millets, Fiber crops, Pulses, Medicinal, Herbal and Aromatic plants and Spices and Condiments.

India has traditionally been a country of organic agriculture, but the growth of modern scientific, input intensive agriculture has pushed it to wall. But with the increasing awareness about the safety and quality of foods, long term sustainability of the system and accumulating evidences of being equally productive, the organic farming has emerged as an alternative system of farming which not only address the quality and sustainability concerns, but also ensures a debt free, profitable livelihood option.

Global Scenario of Organic Agriculture

The organic agriculture is developing rapidly. Its share of agricultural land and farms continues to grow in many countries. The main results of the latest global survey on certified organic farming are summarized below:

Growing area under certified organic agriculture

- ☆ 35 million hectares of agricultural land are managed organically by almost 1.4 million producers.
- ☆ The regions with the largest areas of organically managed agricultural land are Oceania (12.1 million hectares), Europe (8.2 million hectares) and Latin America (8.1 million hectares). The countries with the most organic agricultural land are Australia, Argentina and China.
- ☆ The highest shares of organically managed agricultural land are in the Falkland Islands (36.9 percent), Liechtenstein (29.8 percent) and Austria (15.9 percent).
- ☆ The countries with the highest numbers of producers are India (340'000 producers), Uganda (180'000) and Mexico (130'000). More than one third of organic producers are in Africa.
- ☆ On a global level, the organic agricultural land area increased in all regions.
- ☆ About one-third of the world's organically managed agricultural land is located in developing countries. Most of this land is in Latin America, with Asia and Africa in second and third place.

The countries with the largest area under organic management are Argentina, China and Brazil.

- ☆ 31 million hectares are organic wild collection areas and land for bee keeping.

Almost two-thirds of the agricultural land under organic management is grassland. The cropped area (arable land and permanent crops) constitutes a quarter of the organic agricultural land.

Continent wise growth:

- ☆ **Africa** - In Africa, there are almost than 900'000 hectares of certified organic agricultural land. This constitutes about 2.5 per cent of the world's organic agricultural land. The countries with the most organic land are Uganda, Tunisia, and Ethiopia.
- ☆ **Asia** - The total organic agricultural area in Asia constitutes about nine per cent of the world's organic agricultural land. The leading countries by area are China and India. Organic wild collection areas play a major role in India and China, while Aquaculture is important in China, Bangladesh and Thailand.
- ☆ **Europe** – About 23 per cent of the world's organic land is in Europe. About 4.3 per cent of the agricultural area in the European Union is organic.
- ☆ **Latin America** - Latin America constitutes 23 per cent of the world's organic land. The leading countries are Argentina, Brazil, and Uruguay.
- ☆ **North America** - Seven per cent of the world's organic agricultural land is in North America. In North America, almost 2.5 million hectares are managed organically, representing approximately 0.6 per cent of the total agricultural area.
- ☆ **Oceania** - This region includes Australia, New Zealand, and island states like Fiji, Papua, New Guinea, Tonga and Vanuatu. This constitutes 2.8 per cent of the agricultural land in the area and 35 per cent of the world's organic land.

Global Market

Consumer demand for organic products is concentrated in North America and Europe; these two regions comprise 97 per cent of global revenues. Asia, Latin America and Australia are important producers and exporters of organic foods.

Standards and Regulations

In 2009, there were several major developments in the field of standards and regulations. The new EU regulation on organic production came into force as well as the Canadian organic standard. Furthermore, the Australian domestic organic standard was implemented. Canada and the U.S. concluded the world's first fully reciprocal agreement between regulated organic systems, and the EU introduced procedures for approving certification bodies from outside the EU. In 2009, FAO,

IFOAM and UNCTAD started the Global Organic Market Access (GOMA) project. The aim of GOMA is to facilitate equivalence, harmonization and other types of cooperation in order to simplify the process for trade flow of products among the various organic guarantee systems.

There has been modest growth in the number of certification bodies. A growing number of organic producers are certified through Participatory Guarantee Systems (PGS) across the world. PGS are locally focused quality assurance systems. The leading countries with regards to PGS are located in the global South. Several organic standard setters have also developed draft standards for climate "add-ons" for organic certification, and it is expected that the use of carbon labeling by retailers will grow considerably in the future.

Chapter 15

Overview of Organic Farming

Organic farming works in accord with environment rather than against it. This farming system is realistic suggestion for sustainable agriculture if ample attention is paid to this matter. Global organic farming holds implausible potential for climate change alleviation and adaptation, rural progress and livelihood development, food security, and food sovereignty.

There is abundant scope for escalating the competence under organic farms. The organic farming movement in India is rapidly growing. Local demand for organic food is increasing. India is poised for faster growth with growing household market. Accomplishment of organic movement in India depends upon the expansion of its own domestic markets. India has traditionally been a country of organic agriculture, but the growth of contemporary scientific input concentrated agriculture has pushed the barrier. But with the growing awareness about the safety and excellence of foods, long term sustainability of the method and accumulating evidences of being similarly productive, the organic farming has emerged as an substitute system of farming which not only address the quality and sustainability concerns, but also ensures a debt free, lucrative livelihood option.

Organic farming is critical to the success of this nation and meeting our nation's most burning needs. Increasing the responsibility of organic farming presents the following challenges:

- Implementing policy changes that ensure organic farming meets the rapidly growing consumer demand.
- Rising public responsiveness of the value of organic farming.
- Conducting research to arrive at essential technological advances.

Introduction to more trainings as well as increase in technical supervision would improve the productivity and efficiency of organic farms in India. Government support and investment is particularly vital for the success of the Indian organic farming progress. The responsibility of the Government is important in inspiring the farmers towards organic farming in the country. Some of the major suggestions for extension of organic farming are:

- ✰ Formation of separate 'green channels' for promotion of organic foods.
- ✰ Declaration of premium prices for organic staple food crops.
- ✰ Creation of demand by more awareness programs.
- ✰ Input/conversion subsidies for encouraging organic growers.
- ✰ More R&D funds on organic farming.
- ✰ Inexpensive and fast certification process *etc.*

There is urgent need to engage more and more scientists to identify the thrust area of research for the progress of eco-friendly production technology. While the findings of further research may support organic farming, or dishonor it, organic farming must be taken sincerely. It is increasingly acknowledged that there can be nutritionally relevant composition differences between organic and conventional foods and there are a number of confirmations for potential benefits of organic food consumption from human cohort studies. However, considerable ambiguity/ controversy remain on whether or to what extent these composition differences influence human health.

Besides, studies that allow a more mechanistic understanding of how organic food consumption does influence health are necessary. This could be based on dietary intervention studies with animal models prone to specific diseases. Still, the impact on human health of the actual low-level pesticide exposure from traditionally produced foods is not obvious. Some studies indicate superior nutritional profiles in organic foods than in conventional foods, but the differences are mostly minute and may not be of realistic relevance in well-nourished populations.

To overcome this doubt it is indispensable to:

1) Address a range of practical issues in both the existing meta-analyses of composition data and dietary cohort studies
2) Carry out additional well-designed food composition comparisons for definite crops and meat types to permit reliable comparisons of dietary intakes of nutritionally relevant compounds with organic and conventional foods
3) Carry out well-designed human nutritional intervention studies comparing the effect of organic vs. conventional food consumption on health and health-related physiological parameters

In summary, it is time that the many benefits of organic agriculture are acknowledged by the public and policymakers. A novel combined policy to support organic farmers and the organic food industry is needed. The government should increase research that is responsive to organic farmers' requirements, put together organic programs into each federally funded institution of higher education, and create farm security and assistance programs that work for organic producers.

Index

P

Q

R

S